T0132688

Applications of Plasma Technologies to Material Processing

Applications of Plasma Technologies to Material Processing

By

Giorgio Speranza
Fondazione Bruno Kessler, Trento, Italy

Wei Liu
Fondazione Bruno Kessler, Trento, Italy

Luca Minati
Immagina Biotechnology, Trento, Italy

CRC Press
Taylor & Francis Group
Boca Raton London New York

CRC Press is an imprint of the
Taylor & Francis Group, an **Informa** business

CRC Press
Taylor & Francis Group
6000 Broken Sound Parkway NW, Suite 300
Boca Raton, FL 33487-2742

© 2019 by Taylor & Francis Group, LLC
CRC Press is an imprint of Taylor & Francis Group, an Informa business

No claim to original U.S. Government works

Printed on acid-free paper

International Standard Book Number-13: 978-0-367-20980-3 (Hardback)

Library of Congress Cataloging-in-Publication Data

Names: Speranza, Giorgio, author. | Liu, Wei (Chemical engineer), author. | Minati, Luca, author.
Title: Applications of plasma technologies to material processing / by Giorgio Speranza, Fondazione Bruno Kessler, Trento, Italy, Wei Liu, Fondazione Bruno Kessler, Trento, Italy, Luca Minati, Immagina Biotechnology, Trento, Italy.
Description: Boca Raton, FL : CRC Press, Taylor & Francis Group, [2019] | Includes bibliographical references and index.
Identifiers: LCCN 2018060437| ISBN 9780367209803 (hardback : alk. paper) | ISBN 9780429264658 (ebook)
Subjects: LCSH: Plasma engineering. | Materials.
Classification: LCC TA2020 .S64 2019 | DDC 621.044--dc23
LC record available at https://lccn.loc.gov/2018060437

Visit the Taylor & Francis Web site at
http://www.taylorandfrancis.com

and the CRC Press Web site at
http://www.crcpress.com

Contents

CHAPTER 1

Introduction

S INCE THE 1980S, GASES made of ionized particles have been used to synthesize materials with a specific desired surface chemistry or modify the chemical surface properties of an object. Such an ionized gas is called plasma and is usually referred to as the fourth state of matter in addition to the conventional solid, liquid, and gaseous forms. There is a rather broad variety of plasmas and they can be classified on the basis of different criteria, as summarized in Table 1.1.

Low pressure and atmospheric plasmas can be distinguished on the basis of the operative pressure. In the first case, plasmas are generated in vacuum chambers where the precursor pressure typically ranges between a fraction of a millibar and 10^{-4} millibar. Atmospheric pressure plasma jets operate at ambient pressure without the need for complex equipment to reach low pressures. Due to the combination of simplicity, low cost, and wide possibilities of material treatment and modification, at present they are one of the most promising technologies. Plasmas can be classified also on the basis of temperature, that is, thermal equilibrium. In non-thermal plasmas, the temperatures of the electrons and the ions are in a thermal non-equilibrium. Due to the different mass, the temperature of the electrons (i.e., the kinetic energy) ranges

TABLE 1.1   Classification of Plasmas Following Different Criteria

Criteria	Plasma
Operating pressure	low-pressure plasma
	atmospheric-pressure plasma
Thermodynamic equilibrium	thermal or equilibrium plasma ($T_{electron} \approx T_{ion} \approx T_{gas}$)
	non-thermal plasma or non-equilibrium plasma ($T_{electron} >> T_{ion} \approx T_{gas}$)
	low-temperature plasma ($T_{gas} < 2000$ K)
	high-temperature plasma ($T_{gas} > 2000$ K)
Plasma generation	microwave frequency discharge (300 MHz–300 GHz)
	radio frequency discharge (450 kHz–3.0 MHz; 13.56 MHz)
	DC discharge
	dielectric barrier discharge
	corona discharge
	electric arc
	hollow cathode discharge
	electron beam
	plasma torch
	alternating current
Type of coupling	inductive coupling
	capacitive coupling

between several electron volt (eV), whereas the temperature of the positively charged ions and neutral species is around room temperature.[1,2] This corresponds to a quite low overall plasma temperature ranging from 300 K to 1000 K. For this reason, the plasma is also called "cold plasma" and is favorable for the synthesis of materials at low temperature. In high-temperature plasmas, the temperatures are about 10^7 K, which is typical of fusion plasmas[3] used to produce energy. We will focus on low-temperature, plasmas, which are used to modify the properties of materials.

The common feature of all the plasmas is that they use vaporized chemical compounds or gaseous precursors. In a typical plasma process, the precursor molecules are introduced and ionized in a chamber (the plasma reactor) with selected concentrations. Then, another kind of classification may be done on the basis of the energy source used to generate and sustain the plasma. The source of energy can be thermal energy, a flame, a laser, a microwave or radio frequency (RF), an alternating electric/magnetic field, a dielectric barrier discharge, or electric arcs. At equilibrium, the rate of ionization should correspond to the rate loss of charged particles through recombination of ionized species with electrons or through diffusion toward the boundaries.

The ionized individual atoms or molecules collide, and this triggers the formation of particles by a homogeneous nucleation process. The deposition of thin films or the production of nanostructures on a substrates proceeds through nucleation and coalescence processes which are well described and understood for standard plasma processes.[4–11] By ongoing collisions of particles and by coagulation, nanostructures are formed, and coalescence in thin films is obtained, increasing the deposition time. The utility of plasma processes arises from the significant intensification of chemical processes at the substrate surface, and often promotion of chemical reactions impossible in conventional chemistry.[12]

Thermal plasmas are produced at high pressure (>10 kPa) using direct or alternating current or radio frequency or microwave sources. They are known as plasma torches where electron and ion temperatures are of the order of 1–2 eV which, assuming a Maxwell–Boltzmann energy distribution, correspond to temperatures ranging from 7000 K to 15,000 K. Plasma torches have very low gas ionization but high power from KW to MW.[13,14] Due to the high plasma temperature, torches are used to destroy toxic and dangerous substances or, as in the case of the plasma spray, to produce thick coatings.

As mentioned previously, in the case of non-thermal, or nonequilibrium plasmas, the temperatures involved are much lower and are near 300 K. They are produced at low pressure, under vacuum conditions, generally using low-power RF or microwave or DC sources. In these conditions, the radicals present as ionized species in the plasma react with the material surface inducing a chemical modification grafting different functional properties with respect to the bulk material or leading to the deposition of thin films.[14] Finally, in the case of RF energy sources, inductive or capacitive coupling between radiation and ionized precursor species can be used to classify plasmas. This will affect the kind of apparatus we will discuss in more detail later.

Thanks to their high versatility, efficiency, and isotropic nature, plasma-based processes for solid surface modification and film depositions have gained much interest in connection with several industrial applications in recent years.

CHAPTER 2

Plasma Reactors

A PLASMA IS A CONFINED ionized gas whose temperature and pressure may be tuned in a rather wide range; cold or hot plasmas, low- or high-pressure plasmas can be identified, as already pointed out. Plasmas are generated through a separation of electrons from their parent atoms or molecules, which are left in an ionized status. In this section, we will confine our attention to low-pressure plasmas. In this case, the generation and sustaining of low-temperature plasmas are usually performed by applying electric or magnetic fields to a neutral gaseous precursor. The interaction of a neutral gas with high-energy photons due to cosmic radiation or natural radioactive decay induces dissociation of some molecules in ions and electrons. Then, by applying an electric/magnetic field, it is possible to accelerate such ions and electrons. This leads to new dissociation events when these accelerated particles collide with the neutral precursor molecules thus leading to their ionization. Ionized molecules are essentially radicals which may react more easily with the surface of a substrate changing its composition. To obtain higher efficiency to generate and sustain the plasma, electric fields oscillating in the range of radio frequencies are generally applied to gaseous precursors. However, because of the different transmission power of the gaseous precursor (impedance) with respect to that of the source/ electrical cable system, coupling a microwave radio frequency (RF) source to a gas is not immediate. The different impedance causes a fraction of the electric power to be reflected back to the RF generator. To avoid this problem and efficiently transfer the generated power to the gas, a matching box is used to minimize impedance differences and RF reflections, avoiding loss of power. The performances of the plasma instruments depend essentially on the kind of plasma source and the specific application.[15-23]

In Figure 2.1, the schematic configuration of the most popular plasma reactor systems is represented. Figure 2.1a represents a capacitive-coupled plasma. An RF excitation is applied between two electrodes generating an oscillating electric field. One of these electrodes is the sample holder and is generally grounded. In

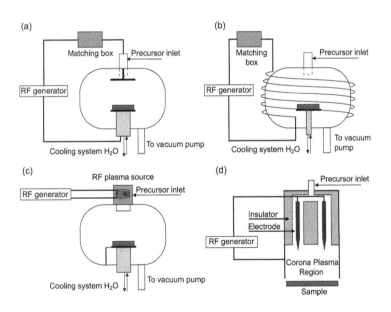

FIGURE 2.1 Schematic representation of the more popular plasma reactor systems: (a) Capacitive-coupled plasma reactor; (b) and (c) inductively coupled plasma reactor; (d) atmospheric plasma reactor. (Reproduced with permission from Minati et al., *Biophys. Chem.* (2017), *229*, 151.)

Figure 2.1b,c, reactors are inductively coupled. In these systems, the plasma is fired and sustained by using an oscillating magnetic field. The latter is generated by a coil powered by a microwave RF generator and placed externally around the reactor chamber (Figure 2.1b) or inside the plasma source (Figure 2.1c). Reactors (b) and (c) differ because in the former the plasma is created in the central region of the reactor chamber, while in the latter the plasma is generated apart from the reactor chamber. In Figure 2.1c, the plasma is formed inside a plasma source and an afterglow propagates inside the reactor till reaching the sample surface. This is a remote plasma as the sample surface is not directly exposed to the exciting RF radiation. In remote plasmas, a low thermal power is transferred to the sample, avoiding sample heating while maintaining efficient surface treatments. For this reason, they are utilized in the treatment of organic materials. Reactors (a), (b), and (c) operate at low pressure obtained connecting the reactor chamber to vacuum pumps. On the other hand, atmospheric plasma may be generated by systems schematized in Figure 2.1d. In these reactors, the gaseous precursors are introduced among electrodes connected at high potential. The intense electric field activates a discharge (the corona discharge). The ions accelerated by the electric force induce the formation of other ions in a cascade process sustaining the plasma. The ionized species are then transported to the sample surface by the precursor gas flow.

Different plasma sources generate different plasmas possessing different ion densities and energies. Then the plasma treatments depend on the plasma operating conditions, the kind of precursor used, the plasma source, the reactor design, and the nature of the material processed.

Plasma treatments lead to five major results:

- Surface cleaning with removal of organic contaminants or microorganisms from the surfaces and surface sterilization

- Etching of the material surface leading to ablation of top layers and surface restructuring with increase of corrugation and surface area

- Crosslinking or branching of superficial molecules causing modification of the surface chemistry

- Grafting of specific functional groups and modification of surface chemical composition

- Coating deposition

The presence of part or all of these effects makes the plasma surface processing suitable for a list of applications summarized in Table 2.1.

When ionized, precursors' fragments accumulate on the substrate surface – the effect of the plasma treatments is the deposition of thin films on the substrate. Film coatings are deposited to add desirable characteristics while preserving the mechanical properties of the substrate. These plasma treatments are suitable to obtain hard coatings,[24,25] biocompatible materials and systems,[26,27] antibacterial materials,[28,29] drug delivery devices,[30,31] transparent conductive coatings,[32–34] next-generation of nanobiointerfaces,[35,36] anti-corrosion coatings, coatings with enhanced tribological properties,[26,37,38] and many others. The plasma processes more commonly utilized to make the depositions are the physical vapor deposition (PVD), the chemical vapor deposition (CVD), and the plasma-enhanced chemical vapor deposition (PECVD). In PVD processes, the coating is obtained by aggregation on a substrate of neutral or ionized atoms/molecules. A target is sputtered or heated until evaporation to produce the ionized or neutral particles.[39,40] In chemical vapor depositions, the production of a thin solid film involves a chemical reaction of vapor-phase precursors. These chemical reactions of precursor molecules may be promoted by heat (thermal CVD), high-frequency radiation such as radio frequency or UV photons (photo-assisted CVD), or by a plasma (plasma-enhanced CVD).[41,42] In the PECVD technique, electrons provided with enough kinetic energy induce dissociation and ionization of the precursor molecules, that is, chemically active radicals, with high efficiency.

TABLE 2.1   List of Plasma Processes and Relative Applications

Surface Modification	Application
Etching	
	surface structuring
	surface cleaning
Functionalization	
	hydrophilization
	hydrophobization
	graftability
	adhesability
	printability
Structural modification	
	diffusion and bonding
	implantation for hardening
Deposition	
	change of hardness
	Young's modulus
	toughness
	change tribological properties
	corrosion protection
	architecturing
	crystal growth (diamond, graphene …)
	scaffolds for biomedicine
Tribological properties	
	wear coefficient
	coefficient of friction
	erosion rate
Optical	
	refractive index, n
	optical loss, absorption coefficient
	extinction coefficient
Electrical	
	circuit breakers
	spark gap switches
Heat	
	welding/cutting arcs
	plasma spray
	thermoelectric drivers

The synthesis of 1D nanoparticles or 2D thin films via plasma involves both gas-phase and surface reactions.[43] These, in turn, depend on the plasma density which can be adjusted varying the total pressure inside the plasma reactor. At pressures of the order of 1 Torr and greater, both heterogeneous and homogeneous pathways are present, while at low pressure ($P < 10^{-4}$ Torr), gas collisions are negligible, and the precursor decomposition becomes strictly heterogeneous. The material growth in these conditions is governed essentially through surface processes, thus enabling a precise control over the material properties.[43,44]

New trends in plasma technology have been triggered by the introduction of new technologies allowing treatments in the spatially controlled micron range (1–1000 μm) in a temporal scale (plasma pulses < 50 ns), plasmas in liquid environments. Figure 2.2 illustrates how the availability of new plasma sources fosters the plasma science research as conceived by the 2017 plasma roadmap.[45]

Apart from ozone generators based on localized microdischarges that nevertheless are distributed randomly in time and space, the commercialization of reproducible low-temperature

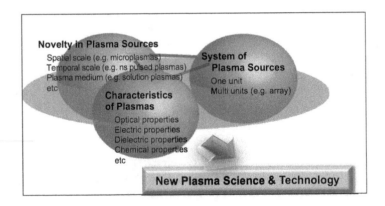

FIGURE 2.2 Diagram illustrating the flow from novel plasma sources to new plasma science and technologies. (Reproduced with permission from Adamovich, I. et al., *J. Phys. Appl. Phys.* (2017), *50* (32), 323001.)

microplasmas generated in cavities is of recent origin. Essentially, they are plasma jets, generated in air or argon, and operating at a power from 5 to 400 W. Possible applications are surface cleaning, medical therapeutics, and cutting for power $\geq$ 100 W. This technology was recently applied to the synthesis of nanostructures obtained by electric discharge microplasmas in supercritical fluids.[46] In supercritical conditions, the microdischarges lead to the bottom-up growth of nanomaterials as illustrated in Figure 2.3a.

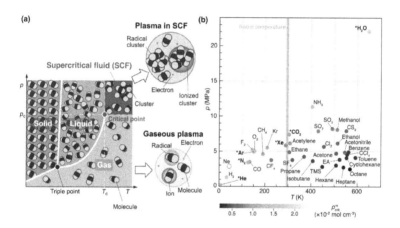

FIGURE 2.3   Schematic of phase diagram and position of critical points as a function of temperature and pressure for various gases, molecular compounds, and organic solvents. (a) Phase diagram of a linear molecule with solid, liquid, gaseous, and supercritical fluid (SCF) phases. The difference in the molecular structure of the four different phases is also indicated. The illustration also shows conceptually the difference between conventional gaseous plasmas and plasmas generated in SCFs, where the high density of the medium leads to the formation of radical and ionized clusters. (b) Critical points in relation to room temperature (between 20 and 26°C), indicated by the gray vertical line. TMS: Tetramethylsilane; EA: Ethyl acetate. The gray scale of the points indicates the molar critical density, ρ^m_{crit}, between 0.2 and 2.4×10^{-2} mol cm^{-3}. The "*" marks substances which electric discharge plasmas have been generated inside the SCF phase. (Reproduced with permission from Stauss et al., *Phys. Plasmas* (2015), *22*, 057103.)

Figure 2.3b displays the critical values (T_{crit}; p_{crit}) of various substances in relation to room temperature (i.e., between 20 and 26°C). At the extremes are He with $T_{crit} = 5.2$ K and $p_{crit} = 0.227$ MPa, and H_2O with $T_{crit} = 647.2$ K and $p_{crit} = 22.06$ MPa. The gray scale indicates the levels of the critical molar density, $\rho^m{}_{crit}$ obtained with discharge plasmas in the indicated supercritical fluids. This technology has been applied successfully to the synthesis of molecular diamonds, carbon films, and carbon nanostructures when operating in CO_2 atmosphere, tungsten, and copper films.[46]

CHAPTER 3

Plasma Applications

$\mathbf{P}$LASMA PROCESSES HAVE BEEN developed to attain a variety of specific surface properties. They can be utilized to remove undesired molecules as in plasma cleaning or to modify the surface chemistry by grafting functional groups. Also, plasma treatments may consist of thin-film depositions needed to deeply modify the surface properties to increase the wear resistance, the chemical inertness, or the electronic properties, as we will show in the following examples.

3.1 PLASMA CLEANING

Cleaning procedures are essential when a high degree of surface quality is required. In this respect, treatments with oxygen plasmas are a safe, efficient, and easy alternative to common cleaning methods. A wide variety of industries utilize plasma treatments to remove organic contaminations from material surfaces needing a high degree of cleaning. They are suitable for the cleaning of circuit boards, the removal of organic contamination from glass slides and flat panels, and the cleaning of medical devices and other materials, such as various metals and ceramics. The active species in plasmas combined with vacuum ultraviolet (UV) radiation energy chemically react with the surface contaminants determining their volatilization and removal. A typical example is the preparation of silicon surfaces for electronics.[47,48] The use of chemical treatments to remove hydrocarbon contaminants deriving from photoresist or other sources always leaves a thin layer of contamination molecules on the substrate surface. Through an appropriate selection of gases and plasma operating conditions, it is possible to remove various materials, to strip photoresists, post-etch, and implant residue, and induce a surface passivation. The gaseous precursor used for cleaning generates the chemically active species which should react with the contaminations to be removed but not with the underlying substrate. For example, an O_2 plasma generates positively and negatively charged oxygen species which can attack the photoresist, leading to a complete decomposition.

Plasma can be utilized also for the sterilization of medical equipment, avoiding the use of chemical disinfectants which are generally toxic not only for bacteria but also for living tissue cells.[49] An even more important problem of current times is antibiotic resistance, which is rising to dangerously high levels in all parts of the world. The complete elimination of bacteria is then becoming an even more important issue. Until now, high temperature (steam, hot metal objects) and disinfectants have been used to kill bacteria in wounds and non-living biomedical articles. A safe, fast, and efficient alternative is offered by glow discharge or plasma generating a reactive chemical atmosphere which is able to clean both living and non-living objects at low temperatures and high rates. The temperature of the ions and the neutral species is low enough to avoid any thermal damage to any object coming in contact with the plasma, while ensuring successful sterilization removing biofilms, bacteria, and fungi.[50–53]

3.2 PLASMA TO ENHANCE ADHESION

Removal of surface contaminations is also used to promote adhesion. Strong interfacial forces via chemical compatibility and/or chemical bonding are needed to induce good adhesion. Then, the composition of the surface and its hydrophilicity are also of practical importance. In this latter respect, hydrophilicity plays a crucial role since it determines the surface interaction with epoxy or similar polar adhesives and then the degree of bondability.[54] Surfaces should be carefully prepared to remove contaminants before bonding. To remove hydrophobic hydrocarbon-based contaminant thin layers, hazardous chemicals are commonly used to clean the surfaces and increase their hydrophilicity.[55,56] The need of less dangerous and toxic and more environmentally friendly methods is then coming forth. Plasma surface treatments are beneficial because they can assist the process of interfacial adhesion, creating chemically active functional groups such as amine, carbonyl, hydroxyl, and carboxyl groups. For this reason, plasma is used to improve bondability on substrates such as glass, polymers, ceramics, and various metals. Plasma treatments were applied to graft oxygen-based functional groups to carbon fiber-reinforced polymers. This led to an improved wettability of the carbon fiber surfaces and roughness, that is, higher surface exposed to the epoxy resins. As a consequence, a significant improvement of the interfacial adhesion between carbon fibers and the epoxy resin was obtained.[57] Carbon fiber-reinforced polymers are largely used in a wide range of engineering applications where they are coupled with metallic fixtures and fittings, in particular those made of titanium and related alloys.[58] Plasma treatments of reinforced polymers resulted in optimal bonding to such systems, improving adhesive properties. Hybrid materials combining thermoplastic polymers in conjunction with metals are increasingly introduced as structural materials for the protection of buildings and infrastructures in the automotive industry. Interactions developing at the interface between a metal and polymer/composite strongly influence the adhesion and the efficacy of the protective coatings. Plasma-activated chemical vapor deposition was used to add silicon, carbon, and hydrogen radicals on the surface of steel with the purpose to create reactive sites that promote the adhesion to thermoplastic polymers such as polypropylene, polypropylene maleic anhydride, and polyurethane.[59] The authors found that the presence of polar groups ($-NH_2$, $-C=O$, and $-OH$) increases the surface energy and adhesion, while those treatments such as the deposition of hydrogenated $-CH$ thin films result in loss of coupling. In general, the adhesion enhancement may be ascribed to an increased density of oxygen-based functionalities, while no particular adhesion benefit is obtained grafting amine functional groups.[60] An important area of plasma applied to adhesion is the treatment of biomaterials. As an example, plasmas are employed in dentistry to improve the resin adhesive penetration, and in particular that of hydrophilic monomers. Again, the use of atmospheric plasma enhances the wettability of dentine surface. The water contact angle of dentine surface decreases from 65° to 4° with 30 s of plasma treatment.[61] A significant increase in adhesive-denting bonding strength was obtained after atmospheric non-thermal argon plasma treatments.[62]

Plasmas are also utilized to improve the integration of metal joining elements. Compared to traditional joining techniques such as welding and riveting, adhesive bonding offers additional advantages such as acoustic insulation, vibration attenuation, structure lightening, corrosion reduction, and uniform stress distribution.[63] Finally, among others, plasma treatments are utilized to improve the adhesion of polymer coatings to metals also for passivation. Atmospheric pressure plasmas (DBD and plasma jet) were used to treat the surface of aluminum alloys, leading to a significant increase in the adhesion of superhydrophobic[64] coatings and passivating coatings.[65]

3.3 PLASMA FOR FOOD AND PACKAGING

A novel application of cold plasma technology relates to the preservation of food. The non-thermal plasmas are highly advantageous for their efficiency in the decontamination of food products from bacteria and spoilage/pathogenic microorganisms thanks to the huge amount of reactive oxygen species generated.[66-69]

To reduce the risk of bacterial contamination and food degradation, plasma technology is applied also to improve the barrier properties of packaging.[70,71] However, the efficacy of the plasma treatments strongly depends on the process parameter used. The nature of the reactive radicals depends on the kind of gaseous precursor utilized, while the operating conditions of the plasma source (e.g., frequency and input voltage) determine their density and energy, which in turn are linked to the treatment effectiveness.[72] As for the precursors, oxygen and nitrogen plasmas produce a large number of radicals including NO_2, NO, OH, O_3, O_2, O. These active species react with the double bond of radical-sensitive unsaturated fatty acid inducing oxidative stress[73] and membrane damage. During the plasma treatments, the cell membrane is exposed to intense electric fields which can cause membrane rupture[74] The resulting altered permeability was found to be the most likely mechanism leading to cell death. Another important aspect is the kind of microorganisms to be killed because their inherent properties determine their sensitivity to plasmas, which differs within species or even strains.[75,76] Experimental results show that sporulated bacteria and the presence of chitin cell walls increase the resistance to plasmas with respect to vegetative cells.[77] Cold plasmas can be applied to improve the properties of materials for packaging. Grafting different functional groups on packaging material modifies its surface energies to enhance antimicrobial and mechanical properties (glazing, adhesion, sealability, wettability, dye uptake, etc.).

The increase of surface energy has been utilized for decades to improve the printability of packaging polymers.[78] Recently, several solutions have been proposed and used to develop antimicrobial packaging. They include: (i) the presence of pads containing volatile antimicrobial compounds; (ii) the incorporation of antimicrobial compounds into the structure of the packaging material; (iii) the adsorption of the antimicrobial compound in coatings or directly on the surfaces of the polymers; (iv) the immobilization of antimicrobial agents on the material surface by ion or covalent bonding; and (v) the application of antimicrobial substances such as chitosan.[79] All these solutions profit from plasma treatment, which renders the next surface manipulation more efficient. Plasma treatments are used to deposit coatings acting as gas permeation barriers. Among them, diamond-like carbon coatings were successfully utilized as polyethylene terephthalate coatings for their chemical inertness, low-friction coefficient, increased hardness, excellent gas barrier, and biocompatibility.[80,81] Deposition of active inorganic coatings (silver and titania nanoparticles) was utilized against microbial proliferation on packaging polymers.[82,83] Plasma-assisted atomic layer deposition (ALD) is a technology to deposit ultrathin conformal layers on any kind of substrates, even those characterized by very high aspect ratios.[84] ALD was utilized to deposit gas barrier coatings mainly made of aluminum oxide on various polymeric substrates to diminish their permeation.[85,86]

A final note is dedicated to the in-packaging processing[87] schematically represented in Figure 3.1. First, the food is placed in a plastic box (or occasionally, glass) in air or a modified gas mixture and sealed. Then it is placed between two external electrodes generating a strong electric field for a short period of time, which causes a plasma discharge inside the package thus leading to decontamination, toxin degradation, and enzyme denaturation with the extension of shelf life of agricultural and meat products.

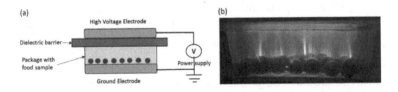

FIGURE 3.1 (a) Schematic of the in-package cold plasma apparatus; (b) plasma discharges in cherry tomatoes. (Reproduced with permission from Pankaj S.K. et al., *Reference Module in Food Sci.* (2017), Chapter 1.)

3.4 PLASMA DEPOSITION OF HARD COATINGS

One of the important areas of application of the plasma technology is the deposition of hard coatings.

There is variety of reasons for the increasing importance of hard and wear-resistant coatings. They are used in a multiplicity of machine operations inducing friction and high temperatures. This causes rapid consumption due to constant mechanical action requiring an increase of the tool resistance. Hard coatings are deposited on moving mechanical parts, for example in cutting, milling, drilling operations, to maintain low-contact friction of sliding and rolling contacts acting as solid lubricants, on tool tips and/or on machine surfaces as a protective measure and for thermal management. Extreme hardness can be obtained with materials exhibiting grain sizes of a few nanometers. These coatings show hardness values which rival or even surpass those of diamond.[88] Hard coatings composed by nano-grains may be obtained by thermal plasma jets. This technology has peculiar advantages compared to other techniques, including high precursor flow rates leading to high deposition rates, the possibility to use a wide selection of precursors, and the possibility to operate at very high temperatures, which can vaporize even solid particles.[88] A drawback of this technology is that the high temperatures involved, prevent the deposition of hard coatings on soft polymeric materials. The thermal plasma jet technology was applied to deposit boron carbide films.[89] Boron carbide is a very hard material, compared to diamond; it is more stable at higher temperatures, and it also exhibits better chemical stability when in contact with ferrous materials. With respect to cubic boron nitride, boron carbide coatings are characterized by lower intrinsic stresses and have better adhesion to substrates of interest. Finally, compared to traditional deposition processes such as sputtering, thermal plasma jets allow a high deposition rate of several hard ceramic materials with a nanosize grain structure, which makes boron carbide an attractive coating material for wear-resistant applications.[88]

The same technology may be applied to deposit silicon carbide and titanium carbide.

Thanks to the low density, the high strength, the low coefficient of thermal expansion, the high thermal conductivity and good electrical properties, the high decomposition temperature (2498°C) and the excellent resistance to corrosion (i.e., alkaline and acid solutions), silicon carbide is used for different applications. As for titanium carbide, it is a super-hard, heat-resistant, high-melting material, widely used for manufacturing metal tools and protective coatings and carbide steel.[90] Besides high temperature and wear stability, the chemical composition of the coating is important for hard passivating coatings. For example, to avoid oxidation of the coating at high temperatures, Al, Cr, Si, B, and other oxide- and nitride-forming elements were introduced in the protective surface layers. Single-phase TiN-TiAlN coatings were deposited by ion-plasma techniques for hardening the surfaces of the rolling tools.[91] Besides TiN and TiAlN, other compositions based on titanium nitrides are utilized for resistant hard coatings, encompassing TiAlCrN, TiAlBN, TiAlSiN, TiAlBSiN, TiN-Si_3N_4, TiN-TiB_2 and other examples.[92-97] Other high-tech coatings capable to form high-temperature lubricating films based on vanadium, molybdenum, and other Magnéli-phase oxides[98-100] are being researched. Molybdenum, carbon tungsten-based coatings are utilized for the so-called *temperature adaptive chameleon behavior*[101] schematically represented in Figure 3.2. This process involves complex adaptation of the chemical-physical and structural properties of the contact surfaces. The changes in these properties self-adapt toward low-friction and low-wear contact conditions. In particular, a functional coating is designed to provide the necessary resistance to the loads and fatigue and mechanical integrity similar to that of hard coatings. This is achieved with a functional "chameleon" interface (see Figure 3.2a) encapsulating hard crystallites in a matrix acting as a diffusion barrier, which prevents degradation of the functional coating. Also, solid lubricant elements are added to

(a)

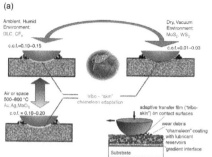

(b)

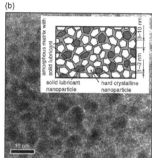

FIGURE 3.2 (a) Conceptual design of chameleon or smart solid lubricating coatings. The scheme shows a tough nanocrystalline/amorphous coating with a functionally gradient interface. (b) TEM image showing nanocomposite structure of an yttrium-stabilized zirconia–Au–MoS$_2$–DLC coating. Inset schematizes the chameleon-like nanocomposite coating structure. (Reprinted with permission from Voevodin et al., *Encyclopedia of Tribology* (2013) Springer.)

the matrix to minimize the loss of composite hardness and elastic modulus. Recent strategies were applied to achieve this adaptive behavior[102]: (1) temperature-activated diffusion of metal lubricants to the surface; (2) temperature- and environment-activated formation of lubricious oxide phases; (3) temperature- and strain-actuated structural evolutions in the contact.

Examples of these structures are generated using yttrium-stabilized zirconia and gold nanocomposite. Figure 3.2b shows a TEM image from a YSZ–Au–MoS$_2$–DLC chameleon interface and its nanostructures. The inset represents a schematic of the interface composition. These coating materials demonstrate high hardness, but at the same time, the presence of Au induces high ductility and fracture resistance. Moreover, when heated to temperatures of 500°C, they provide a low-shear interface on a hard surface.[103] Other studies use Ag as noble metal inclusion embedded in ceramic matrices,[102] or multilayered structures. As an example, Ag multilayered coatings are made by two adaptive Y-stabilized

Zr as hard coating, an Ag-Mo lubricant layer, and TiN as a diffusion barrier between them.[104]

3.5 PLASMA DEPOSITION OF PASSIVATING ANTI-CORROSION COATINGS

Anti-corrosion protecting layers are one of the most important areas of coating application both in terms of total coated surface areas and business volume. It has been estimated that each year the cost caused by corrosion damage is ~3–4% of the gross domestic product of a country.[105] It is easy to calculate that "corrosion costs" for economically strong countries are in the order of $100,000 million, and for the whole world (estimated for 2012) they are ~$2,151–$2,868 billion.[106] Therefore, improving the durability of materials preventing corrosion has an important impact on country economies. Corrosion protection is generally accomplished through a passivating coating which has to be deposited on the material's surface. Generally, pretreatments are needed to prepare the surface before coating application. A common problem affecting the protective coatings is the presence of nanoscale defects (inhomogeneities such as columnar structure), of microscale defects (voids, flake, over-coated particles, pinholes), or macroscale defects (scratches, pits, grooves, ridges). When exposed to a corrosive environment, these may lead to serious corrosion attacks in spite of the high passivating resistance of the coating.[107] Plasma deposition techniques may be the solution to this problem. Anti-corrosion films may be deposited by PVD and plasma-assisted chemical vapor deposition (PACVD) leading to a broad range of structural properties. To optimize passivating structures, zone models relating to the microstructure of a thin film, to the T_S/T_m (T_S: substrate temperature, T_m: melting temperature of the coating material), to the total pressure, and to the kinetic energy of the condensing species[108] are used. Dense, low-porous coatings are observed for both higher substrate temperatures and kinetic energy of the condensing species. When the substrate temperature cannot be increased, high kinetic energy of the condensing species may partially solve the problem.[106] In relation to the kind of substrate to be protected, a number of deposition methods may be chosen to optimize the coating properties. To improve the properties of the corrosion barriers, a number of deposition methods may be selected that are as follows: (i) cathodic arc evaporation; (ii) pulsed laser deposition; (iii) ion beam-assisted deposition; (iv) inductive coupled plasma-assisted magnetron sputtering; (v) electron cyclotron resonance magnetron sputtering; (vi) hollow cathode magnetron sputtering discharges; or (vii) high-power pulsed magnetron sputtering.[109–113]

Among the coating properties, the thickness significantly influences the corrosion behavior.[114] Corrosion and tribological properties of a-C:H:Si films deposited on metallic surfaces were compared with diamond-like and chromium plating.[115] The authors found an increased load capacity with higher coating thickness in the a-C:H:Si. These films outperform diamond-like carbon films, which are characterized by high chemical inertness but high stresses when the thickness increases, thus leading to the presence of defects.[115] On the other hand, films based on a-C:H:Si show that increased thickness distributes the stress more homogeneously during loading so that nearly no stress appears at the interface. This results in longer coating durability with respect to the DLC films. The a-C:H:Si show better wear resistance with absence of delamination and a higher corrosion potential as well as a far lower current density if compared to hard chromium films. A similar work was carried out by other authors depositing amorphous films such as hydrogenated silicon carbide (a-SiC$_x$:H), silicon nitride (a-SiN$_x$:H), and silicon carbonitride (a-SiC$_x$N$_y$:H) protective coatings on stainless steel.[116] They found that 20-nm-thick films were unable to avoid stainless steel corrosion while good passivation was obtained with 100-nm-thick layers. Essentially, increasing the barrier's thickness reduces the probability of pits and deep porosity which make the substrate accessible to external harsh elements. However, depending on the kind of material deposited, thickness easily leads to more stressed coatings which could detach from the substrate. A possible solution to this problem is the use of plasma-enhanced atomic layer deposition. Using this technology, zirconium oxynitride (Zr_2N_2O) films were deposited on a stainless-steel membrane of a polymer electrolyte membrane fuel cells.[117] A thickness of 36 nm of zirconium oxynitride films is enough to obtain good passivation properties, while in the case of ZrN, good corrosion resistance needs 10-μm-thick films. ALD deposition while incorporating a controlled dose of oxygen atoms into ZrN by PEALD leads to a lower interfacial contact resistance, which is desirable for a better stack performance. This kind of coating is

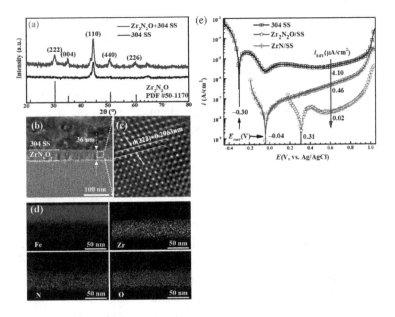

FIGURE 3.3   (a) XRD; (b, c) cross-sectional FIB-TEM images; (d) EDS elemental mappings (Fe, Zr, N, and O) of the ZrN_xO_y-coated 304 stainless steel; (e) potentiodynamic polarization curves of virgin 304 stainless steel, ZrN_xO_y-coated 304 stainless steel, ZrN-coated 304 stainless steel. (Reproduced with permission from Wang X.Z. et al., *J. Pow. Sources* (2018), *397*, 32.)

displayed in Figure 3.3. The X-diffraction spectra in Figure 3.3a shows the presence of crystalline phases in Zr_2N_2O, which is further proven by TEM images (Figure 3.3c). The coating structure is illustrated in Figure 3.3b, while the element distribution is shown in Figure 3.3d. Finally, electrical measurements indicate that Zr_2N_2O films exhibit an electrical conductivity similar to that of ZrN but with higher corrosion potential, leading to the remarkable corrosion resistance of ZrO_2, as shown in Figure 3.3e. Several works have been published in literature concerning the protective properties of TiN coatings. Barshilia et al. studied the protective properties of single TiN and TiN/NbN multilayer coatings deposited by reactive DC magnetron sputtering process.[118] Coatings were deposited on stainless steel with thickness ranging from 32 to 200 nm. They found that the corrosion current density decreased with the film thickness. Similar results were obtained in other works, where the passivating properties of TiN[119,120] and TiAlN, were studied.[121]

To replace environmentally unfriendly hard chromium coatings, CrN/NbN superlattice coatings deposited on 304L stainless steel by magnetron sputtering, cathodic arc evaporation, and arc bond sputtering were used.[122] The steered cathodic and arc bond depositions lead to a highly dense and void-free coating. In contrast, the CrN/NbN coatings deposited by magnetron sputtering show a porous columnar morphology. As a consequence, the passivation properties of the coatings grown by arc bond technique were superior to those grown by either UBM sputtering or arc evaporation with clear passivation behavior and low corrosion current densities of 10^{-8} A·cm^{-2}.

Other authors, notably Lv et al., deposited CrAlN films on stainless-steel substrates by mid-frequency pulsed magnetron sputtering.[123] The substrate temperature was maintained at 100°C while different bias voltages (−50, −100, −150, −200, and −250 V) were utilized for the deposition. A columnar structure was obtained at the lower voltages while at −250 V a dense glass-like microstructure is observed. Consequently, higher anti-corrosion properties were obtained with increasing bias voltage.

3.6 PLASMA TREATMENTS OF TEXTILES

The growing demand of technical textiles possessing novel functionalities as well as of environmentally friendly production processes led to an increasing interest in the surface modification and coating processing of textiles. The textile industry transforms various fibers first into yarn and then into fabric-related products. Textile dyeing and finishing are needed to commercialize the products. The dyeing and printing processes utilize large amounts of water for wetting the fibers and dissolving the dyes.[124] Since many dyes produce toxic wastewater, treatment based on advanced technologies to remove pollutants from effluent from textile industries are needed.[125] Then, these production routes pose some issues regarding environmental pollution. Generally, water-based finishing processes are applied to the textile, which therefore needs to be dried, and removing water is energy intensive and expensive. Another problem is the massive water consumption of textile, which makes even more important the concern about the use of clean water and the scarcity of water resources for the sustenance of life.[124] In contrast to conventional wet finishing methods, plasma treatments are a dry process, and desiccation can be avoided, with cost reduction. Plasma technology also solves any problem related to water consumption and pollution and as such is, in principle, a very appealing alternative to traditional textile processing. However, the need to work in vacuum conditions at low pressure in a big vacuum chamber with continuous and fast roll-to-roll textile treatments makes plasma technology rather complex. In addition, other problems could arise leading to treatment failure, for example fibers cleaning could have substantial effects on the treatments. Because plasma treatments affect only the surface topmost layers, presence of contaminations or peculiar surface conditions such as weft and warp direction could significantly limit the plasma efficiency.[126] In fact, the 3D structure of the textiles prevents the charged species of the plasma to deeply penetrate and treat the fabric as the wet process does. Plasma conditions, and in particular pressure, are crucial parameters influencing the treatment efficiency, also considering the large surface area generated by the ensemble of individual fibers composing the textile.

An alternative to low-pressure plasmas is those operating at atmospheric pressure thus not needing expensive vacuum equipment and simplifying the continuous and uniform processing of fibers. The corona discharge is the first example of plasma technology applied to textiles which are passing between electrodes connected to high voltage oscillating at low frequency. Because the discharge energy density falls rapidly with distance, a very small interelectrode spacing of ≈1 mm, incompatible with thick fabrics, is required.[126] In addition, corona plasma treatments induce an increase of surface roughness, decreasing fiber cleaning, and suffer from non-uniformity because they mainly affect superficial fibers, being unable to penetrate deeply into the fabrics.[127]

A valid alternative is the dielectric barrier discharge (DBD), which is another technology operating at atmospheric pressure. DBD has been attracting increasing interest thanks to its high efficiency and the possibility of scalability to industrial systems.[128] A DBD reactor is formed by two electrodes, one of which is covered by a dielectric layer that accumulates the transported charge on its surface with the effect of limiting the amount of charge in a single microdischarge and making them more uniform.[128] Another convenient technology is the atmospheric plasma glow discharge (APGD), which is performed by applying a radio frequency source to two parallel electrodes and a potential up to some kV to sustain the discharge. The plasma is generated in a controlled atmosphere, such as helium or argon at atmospheric pressure, providing good performance treatments in a few seconds.[129] Finally, plasma jets, with respect to the other technologies, are more flexible and can be easily applied to any kind of surface shape. Jets are obtained by first applying a high voltage between electrodes immersed in the gaseous precursor to induce ionization and then forcing it to flow through a nozzle, leading to the formation of plasma plume. The plasma jet is able to penetrate porous structures, such as that of textiles, throughout the entire textile volume, thus making for homogeneous treatments.[130]

Plasma treatments of fabrics are applied to improve wettability and hydrophobicity. Several studies have been carried out to increase the wettability of silk,[131,132] polyesters, polyethylene (PE),[133] polypropylene (PP),[134] polytetrafluoroethylene (PTFE),[135,136] cotton,[137-139] wool,[140,141] polyamide (PA),[142] polyimide (PI),[143] and polyethylene terephthalate (PET).[144,145] Carbon fibers utilized to improve mechanical as well as the electrical and electrochemical properties of the material, were treated with the same kind of functionalization used to induce adhesion of reduced graphene oxide to fabric fibers.[146-148] Surface modifications of these polymers by grafting of –COOH, –OH, and –NH$_2$[149] are unquestionably the most studied over the last years. They are obtained using O$_2$ and N$_2$ as plasma precursors. The presence of polar groups on the textile fibers strongly increases their hydrophilicity (for more details, see also Zille et al.).[126] Hydrophilic surfaces are generally desired to improve dyeing by reducing the required dyestuff while enhancing the color's intensity and washing fastness.

To obtain stain-resistant fabrics, generally the hydrophobic character is enhanced by introducing hydrophobic functional groups as coating or graft co-polymerization. In any case, the role of these treatments is to reduce the surface energy of the fabric by adding lower surface energy chemical groups to the surface. PET was successfully modified using SF$_6$ as plasma precursor.[150] Other authors utilized fluorine-based gaseous precursors for treatments such as hexafluoropropylene,[151] fluorodecylacrylate,[152] hexafluoroethane,[153] and tetrafluoromethane.[154] However, manipulation of fluorine compounds is dangerous since losses during manufacturing or disposal are harmful to human health and the environment. Possible alternatives are lauryl-methacrylate for spraying method[155] or hexamethyldisiloxane.[156] In addition, flexible fibers coated with hydrophobic species show frequent surface cracks and delamination, causing the loss of the hydrophobic character.[157] Plasma etching can be a good solution to generate a pronounced hydrophobic character without any drawback of pollution and treatment degradation. Contrary to what was observed in the work by Paosawatyanyong et al., oxygen plasma was used to induce highly hydrophobic surfaces as shown in Figure 3.4.[150] The difference is based on the relaxation of functional groups which tend normally to move under the polymer surface which is then characterized by the presence of unipolar groups only. Usually, this relaxation is slow but can be accelerated by heating the polymer at 130°C, which renders the fibers superhydrophobic. Other kinds of textile treatments include: functionalization carbon fibers[158,159] with silver nanoparticles,[160] in particular, for medical applications to enhance bacteriostatic properties,[161] flame retardancy,[162,163] and photocatalytic cleaning surfaces obtained by adding TiO_2 and ZnO[164,165] nanoparticles to textile fibers.

3.7 PLASMA TREATMENTS FOR BIOMEDICAL APPLICATIONS

The American National Institute of Health describes biomaterials as:

> Any substance or combination of substances, other than drugs, synthetic or natural in origin, which can be used for any period of time, which augments or replaces partially or totally any tissue, organ or function of the body, in order to maintain or improve the quality of life of the individual.

Replacing parts or functions of the human body requires the material to be biocompatible and the degree of biocompatibility is defined on the basis of the kinds of reactions occurring when the material comes in contact with the living matter. The fabrication of biomaterials evolved with time. Over the past few decades, the emphasis shifted from achieving bioinert materials avoiding any tissue response to biomaterials stimulating specific cellular responses. These were called third-generation biomaterials,[166] used to identify bioactive and bioresorbable materials able to

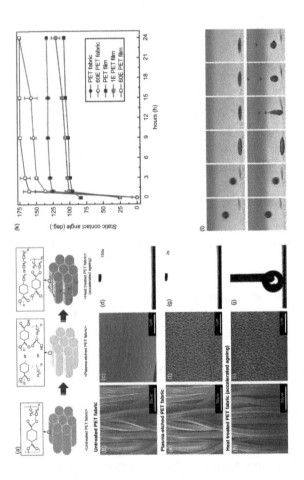

FIGURE 3.4 Schematic of fiber treatment: (a) SEM images and static contact angle change for untreated (b–d), 60 min plasma-etched (e–g), and 60 min plasma-etched and heated at 130°C for 24 h (h–j) PET fabrics; (k) effect of thermal aging time on the contact angles of water drops on PET films or fabrics; untreated PET fabric (filled circle), PET fabric plasma-etched for 60 min and heated at 130°C (empty circle), and untreated PET film (filled square), PET film plasma-etched for 1 min (1E, square filled with cross) and PET film plasma-etched for 60 min and heated at 130°C (empty square); (l) bouncing behavior of water droplets on the PET fabrics untreated (upper strip) and plasma-etched for 60 min and heated at 130°C for 24 h (lower strip). (Reproduced with permission from Oh et al., *RSC Adv.* (2017), 7, 25597.)

activate genes that stimulate a repair of living tissues. The positive tissue reaction is expected to develop as a consequence of the material characteristics imparted through appropriate surface treatments. These involve a surface molecular tailoring to influence the cell microenvironment for achieving specific responses. Recent advances in material science and in the electrophysiological behavior of cells and tissues have inspired the development of a new generation of biomaterials, enabling controlled interactions with the cell functions. Every cell is formed by a plasma membrane where proteins pump ions to generate transmembrane voltage potentials. The possibility to develop fourth-generation biomaterials has been recognized.[167] Specifically, they consist of electronic systems integrated with the body, providing diagnostic and therapeutic functions by delivering bioelectric signals to properly stimulate cell growth and monitor the effect of such stimuli.

Such a complex system requires a perfect integration with the living environment to correctly interact with or control physiological functions. This requires that the surface chemistry be accurately manipulated. Plasma technology coupled with other types of dry/wet surface processing plays an important role in making the surface engineering. Henceforward, we will provide some example of application of plasma technology for bioapplications.

Controlling surface energies is of paramount importance when designing the properties of biomaterials. Hydrophilic and hydrophobic character is deeply at play because it modulates the interaction of a biomaterial with the biological environment. As an example, plasma can be utilized to enhance the performances of a catheter, for example, limiting thrombus formation (vascular catheters) or encrustation (urological catheters). Polymers utilized for catheters include polyethylene, polytetrafluoroethylene, silicone, polyurethanes, and polyvinylchloride.[168] However, these materials show relatively high friction coefficients, which makes the insertion painful. A slip agent is typically applied but it may cause problems associate with debris and pathogen proliferation.

Plasma ion implantation may induce positive surface changes, leading to remarkable reduction in the polyurethane catheter friction coefficient.[168] The reduction in the friction coefficient of the polyurethane surface was achieved by the combined effects of surface corrugation and carbonization. Sterilization is a common problem found in devices that are inserted in the human body, such as catheters. Cold nitrogen plasma treatment was utilized for sterilization and effects evaluated from the bactericidal[169] and surface-aging[170] points of view. Results show that plasma treatments have superior sterilizing properties with respect to conventional methods, while the aging of plasma-treated surfaces shows a fast decay in short times, but after that the surface still retains the presence of functionals after long storage times.[170] Plasma has been successfully applied to functionalize catheter surfaces with silver nanoparticles. The combination of plasma and silver nitrate wet treatments is a fast, easy, and convenient method for obtaining colonization-resistant catheter surfaces.[171,172] More recently, the surfaces of intravascular catheters were plasma-treated to improve the coupling with chitosan coatings, which is known to possess bactericidal properties,[173] or bacterial biofilm-resistant acrylate polymer coatings.[174] Results show that the treated catheter has high biocompatibility and improved bacteriostatic characteristics.

Plasma treatments were utilized also to reduce the formation of thrombi in intravascular catheters. Organosilane precursors were used to deposit a hydrophobic coating on coronary catheters composed of a soft polyether amide block.[175] In another work, heparin-coated Ag nanoparticles were bound to plasma-treated polyethylene catheters, thus obtaining both antibacterial and anticoagulant properties.[176] Adhesion of undesired biological matter is also a common problem found in contact lenses. Plasma treatments may be advantageously applied to reduce the surface biofouling.[177] Biofouling strongly depends on the surface properties such as roughness and chemistry.[178] Generally, an increase of hydrophilicity corresponds to a decrease in fouling surface.[179-181]

Plasma was also used to deposit coatings possessing antifouling properties to stimulate cell adhesion and proliferation.[182–184]

Biomaterials that come in contact with blood or protein require special surface treatments to enhance biocompatibility. Amine functional groups, which are attached by ammonia plasma treatment, act as hooks for anticoagulants, such as heparin, and thereby decrease thrombogenicity.[185–187] Synthetic polymer implant materials can be surface activated using radio frequency plasma techniques to enable covalent immobilization of cell-binding peptides derived from extracellular matrix proteins such as fibronectin.[188,189] Other methods are implemented to obtain improved hemocompatibility. For example, polypropylene fibrous membranes were functionalized grafting a polysulfobetaine methacrylate brush via plasma-induced surface polymerization.[190,191] The treated surfaces display blood-inert and antifouling, anticoagulant, and anti-thrombogenic properties, thanks to the polysulfobetaine methacrylate overall charge neutrality and high hydration capability. In other works, hydrophobic polymers, such as poly(vinylidene fluoride) membranes, were plasma-treated to graft network-like and brush-like PEGylated layers.[192] Grafted peptides can promote complete coverage of a surface with a monolayer of intact, healthy endothelial cells to form a natural, blood compatible surface.[193,194] Surface plasma treatments can play an important role to induce endothelial cell growth providing biocompatible anti-thrombogenic surface properties.

Plasma treatments of biomaterial surfaces are widely utilized in regenerative medicine. It has been found that an appropriate use of plasma treatment can cause higher cellular activities including cell growth, proliferation, and differentiation. Jeon et al. and Sharifi et al. treated electrospun poly(ε-caprolactone) fibers in an oxygen plasma to enhance the density of related functional groups.[195,196] Electrospinning is a convenient technique to produce a mesh composed by nano- and micro-fibers, nanostructured fibers, core-shell fibers, and corrugated fibers as shown in Figure 3.5.

FIGURE 3.5 Different fiber morphologies: (a) beaded; (b) smooth; (c) core-shell; and (d) porous fibers. (Reprinted with permission from Ramakrishna S. et al., *Mater. Today* (2006), *9* (3), 40.)

Hierarchical patterning consisting of a mixture of nano- and micro-sized structures has been studied to understand the effect of the structure architecture on cellular responses. It has been reported that integration of nanoscale structures in a 3D micro-structured scaffold substantially improves their biological performances in terms of tissue regeneration.[197–199] Plasma treatment of poly(ϵ-caprolactone) fibers further improves the adhesion and proliferation of osteoblast-like cells. Improvement of cell adhesion performances was achieved by applying plasma treatments to other kinds of scaffolds such as poly(L-lactide) microfibrous meshes.[200] Another attractive biopolymer is silk fibroin, a natural fiber possessing good biocompatibility and suitable mechanical properties such as elasticity and strength, which can be processed to make rather different kinds of scaffolds for tissue engineering.[201,202] Silk fibroin may also be electrospun[203,204] and plasma treated to enhance the scaffold properties.[205] A combination of electrospinning and plasma polymerization was successfully utilized to produce polyethylene fibers to promote the formation of a stable endothelial layer to avoid platelets adhesion in vascular prostheses.[206]

Plasma treatments were applied to titanium and titanium alloys[207–210] and hydroxyapatite-based scaffolds,[211,212] to mixed polymeric hydroxyapatite scaffolds[213] to induce bone regeneration. Hybrid materials were made infiltrating β-tricalcium phosphate with poly(L-lactide-co-D, L-lactide) (PLA). The same scaffolds were plasma-polymerized with allylamine to improve PLA properties. The effect of plasma clearly appears in Figure 3.6 where only the plasma-treated scaffolds show fibronectin footprints, testifying a higher migration of osteoblast in presence of allylamine.

In research by Bos et al., polystyrene surfaces were treated with a carbon dioxide gas plasma, allowing the immobilization of albumin-heparin conjugates, thus inhibiting the exogenous thrombin.[214] Adding a small amount of fibronectin on top of the albumin-heparin-treated surfaces increases the growth of endothelial cells, which significantly reduces the number of adhering platelets. In work by Dekker et al., the endothelialized bioimplants will have improved biocompatibility and reduced antigenicity and thrombogenicity.[215] Plasma modification of silicone tubes grafting carboxyl and amine functional groups led to enhanced endothelial cell attachment and proliferation, and cells were more stable under fluid shear stress compared to untreated silicon surfaces.[216] Plasma treatments are broadly utilized to stimulate cell proliferation and growth in regenerative medicine, the treatment depending on the kind of substrate and the cell lines to stimulate. For

PLA PLA + PPAAm

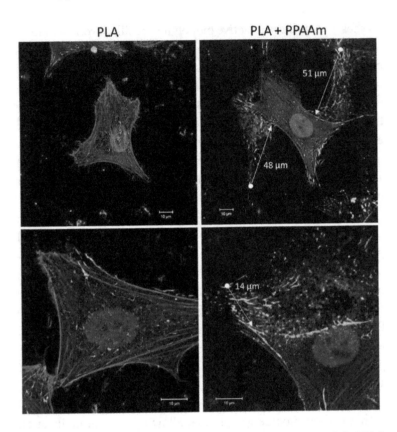

FIGURE 3.6 Cell migration of osteoblasts on allylamine-modified PLA discs (PLA + PPAAm). Note that footprints of fibronectin (bright spots on the right column) were left behind by the migrating cells (arrows mark the covered distances) solely on PPAAm-coated surfaces. In contrast, cells on PLA remained stationary after 48 h of cell culture (left column) (Laser Scanning Microscope 780, Carl Zeiss; bars = 10 μm; lower row: different cell examples magnified). (Reprinted with permission from Bergemann C. et al., *Mater. Sci. Eng.* (2016), *C59*, 514.)

example, in the case of low-density, high-density, and ultra-high molecular weight polyethylene, tests of rat cell adhesion show that argon plasma treatments induce a significant increase in the biocompatibility of the surfaces.[217] For all three samples, Ar plasma caused ablation of the polymers with a dramatic increase in the surface morphology and roughness, thus inducing an enhanced adhesion and proliferation of vascular smooth muscle cells and fibroblasts cells. The same authors showed that the same Ar plasma treatment of polyethylene naphthalate induces surface modifications leading to better performances with respect to polyethylene substrates.[218]

While the effect of surface roughness on cell adhesion has been well established for a long time,[219-222] less clear is the dependence of cell adhesion on surface chemistry. To isolate the effects of the latter, there must be a careful control of the treated surfaces, which must be characterized by the same coarseness. Antonini et al. studied the effect of different plasma deposition of neutral films and films with positive or negative functional groups.[223] The cell adhesion is sensitive to the polarity of the surface charge since the latter led to adhesion of higher/lower amount of extracellular matrix proteins. The kind of surface polarity depends on the kind of proteins which, in turn, depend on the kind of cell line selected for the experiment. The cell needs a support for growth and proliferation. Suitable 3D structures, resembling a natural extracellular environment, are developed for cell growth.[224-229] Generally, the main disadvantage of polymeric substrates is the lack of proper surface properties that stimulate the adhesion and proliferation of cells and differentiate them into the right type of tissue.[230] Again, non-thermal plasma technology offers a valid alternative to wet chemical processing for surface engineering of the different synthetic polymers, such as bio-degradable thermoplastics, for example, poly-L-lactic acid (PLLA), poly-ε-caprolactone (PCL), and poly (ethylene oxide terephthalate)/poly (butylene terephthalate) (PEOT/PBT), which are of interest, as they are relatively cheap, easily manipulated, and exhibit excellent structural properties. A first example of application, traditional scaffolds made of poly-L-lactic acid were treated with an ammonia plasma to enhance cell proliferation.[231] In another work, calcium phosphate scaffolds were treated using an atmospheric pressure plasma jet.[232] Results show an increase in the surface wettability and presence of OH functional groups on the surface, leading to a significant increase in mouse osteoblasts. It was also proven that plasma treatment can induce a better neovascularization.[233] However, care has to be taken in considering these results because as do not transfer immediately in *in vivo* situations,[234] because cell density, distribution, and timing of seeding within the scaffold also contribute to the formation of bone and vessels.[235,236] In other studies, such as the one by Cools et al., it is shown how arc non-thermal helium plasma processing was applied to acrylic acid-coated 3D scaffolds obtained through additive manufacturing.[237]

It is known that inert gas plasmas generate radical sites, causing the incorporation of polar functional groups (plasma activation). Results show that the plasma treatment induces a strong increase in the scaffold wettability caused by an increase in oxygen-based functional groups. The plasma-activated sample is responsible for the highest production of GAG/DNA, thus demonstrating the stimulation of cell activity. Figure 3.7 shows the effect of scaffold plasma treatments on ATDC5 chondroblasts.[237] Localized clustering indicating low affinity of ATDC5 toward the scaffold surface was found for the negative (UNT) and the positive (ITS) controls.

Cross-sectioned methylene blue-stained scaffolds displayed in Figure 3.7a, indicate that chondroblasts were more spread out in UNT and ITS scaffolds than in to plasma activated (PAct) and plasma coated scaffold (PC). This behavior likely depends on the higher hydrophilicity of both plasma-modified samples, which hinders the cell-surface conditioning. However, live dead assays conducted at time point day 1 (Figure 3.7b) on the same scaffolds clearly show that the plasma coated sample possesses the most homogeneous spread of cells and the highest cell density. These conclusions are confirmed by the methylene blue experiments.

3.8 PLASMA TREATMENTS FOR ENERGY

Plasma treatments are widely utilized for energy production and storage because of their unpaired efficiency in tailoring the surface properties of a broad class of materials, in manipulating intrinsic defects, in doping, in modifying the surface chemistry and its

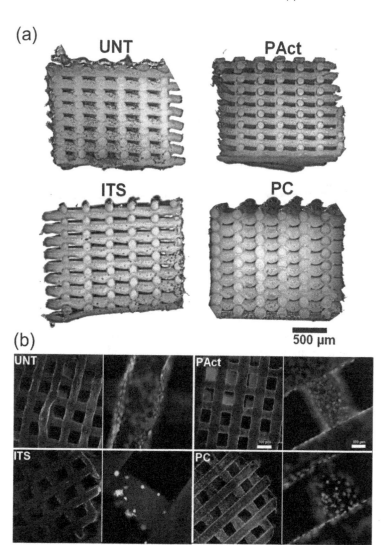

FIGURE 3.7 (a) Methylene blue-stained cross-sections of the scaffolds for the day 1 untreated samples (UNT), plasma-activated samples (PAct), positive control (ITS), and plasma-coated samples (PC); (b) live/dead stained images for the second set of experiments at day 1. (Reproduced with permission from Cools et al., *Sci. Rep.* (2018), *8*, 3830.)

wetting properties, in surface passivation, and in nanostructuration. As a consequence, the plasma technology has a considerable scientific, technological, and also economic–environmental impact, as will be discussed later.

3.8.1 Application in Photovoltaics

Due to environment pollution issues, solar cells are one of the most adopted options for solar energy harvesting. A key issue for such energy devices is the maintenance of optimum performances, which makes the adoption of this technology economically convenient. Degradation of solar cell efficiency may derive from dust and surface fouling. As an example, in desert areas such as Saudi Arabia, dust can cause a reduction in the glass transmission covering the photovoltaic (PV) cells of 70%, and the loss in efficiency can range from 26 to 40%.[238,239] Plasma processing was utilized to nanostructurate glass surfaces, rendering their surface superhydrophobic,[240] preventing grime and ensuring optical transmission/reflection properties thanks to water and dust repellency. The mechanism behind these self-cleaning surfaces mimicking the lotus leaf is obtained by strongly increasing the water contact angle obtained in combination with a reduction of the adhesion. However, unlike nature, a degradation of the superhydrophobic, self-cleaning surface properties has been observed with time.[241] Plasma-based approaches may be coupled with sol-gel techniques to pretreat the substrate in order to enhance the adhesion of the coatings to the substrate and for post-deposition plasma etching of substrates using different fluorinated hydrocarbons to generate surface nanstructures.[242-244] Self-cleaning includes also photocatalytic active surfaces. Coatings based on TiO_2 are deposited by plasma technology as a photocatalyst, capable of breaking down organic contaminants when exposed to light.[245,246] TiO_2 thin layers were deposited on a PET-ITO substrate to block charge recombination processes at the interface between the transparent conductive oxide and the perovskite solar cell, leading to an overall efficiency of 9.2%.[247] In another work, the TiO_2 layer, acting as an electron-blocking layer, was deposited on fluorine-doped SnO_2 glass in a dye-sensitized solar cell.[248] Under 100 mW/cm² illumination, the energy conversion efficiency of the cell was 6.77%.

A higher efficiency of dye-sensitized solar cells was achieved by depositing TiO_2 photoanodes with high-specific surface area and well-aligned TiO_2 nano-dendrites. These nanostructured TiO_2 films led to a photo-conversion efficiency of 12.08% using an N719 dye and iodine ion electrolyte under illumination of simulated AM1.5 solar light (100 mWcm⁻²).[249]

Plasma technology was also utilized to grow nanocrystalline hierarchical nanostructures combining physical vapor deposition of phthalocyanine nanocrystals and plasma-enhanced chemical vacuum deposition of anatase TiO_2 layers and/or nanotubes by controlling the deposition temperature.[250] The plasma-assisted depositions allowed for obtaining a variety of phthalocyanine nanostructures on anatase nanotubes, as displayed in Figure 3.8, and more complex multistacked nanotrees layers offering the possibility to implement these 1D and 3D nanoarchitectures in electronic and optoelectronic devices. These structures were utilized as photoanodes in dye-sensitized solar cells with an efficiency of ~4.6%. Oxide layers are also utilized as antireflective coatings together with amorphous Si-based coatings[251] and fluorides.[252] The conversion efficiency of the photovoltaic modules is limited by losses due to the reflection at air/module interface.[253,252] As a result, design of optimal antireflective coatings to improve solar cell efficiency has gained significant attention by scientists all over the world. Deposition of thin antireflective coatings is produced by various techniques: lithography, template, electrospinning, sol-gel deposition, layer-by-layer, phase separation methods, and plasma-based methods such as chemical vapor deposition and plasma etching.[254] Different kinds of materials deposited as uniform single-layer or structured and patterned films have been developed. They include: porous silicon obtained by plasma etching,[255] silica particles[256] also combined with polymeric materials hybrids,[257] hydrogenated amorphous silicon nitride (a-SiNx:H),

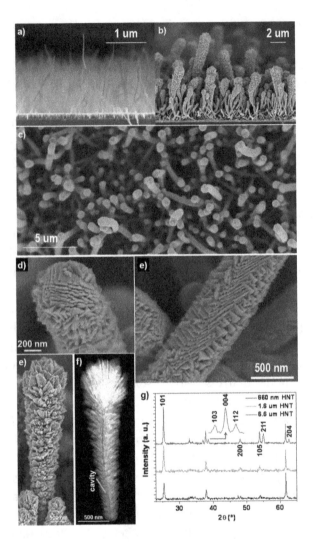

FIGURE 3.8 (a) Characteristic cross-section SEM image showing the formation of phthalocyanine single-crystal organic nanowires using an anatase film as seed layer. (b–c) Cross-section and normal view SEM micrographs of anatase HNT-films respectively of 1.6 μm and 660 nm nominal thickness. Details of a 660-nm thick-HNT tip and lateral appear in (d–e); (e–f) SEM and HAADF-STEM images of a 1.6-um HNT; (g) XRD spectra of HNTs with different thicknesses. (Reproduced with permission from Filippin et al., *Nanoscale* (2017), *9*, 8133.)

aluminum oxide, hydrogenated amorphous silicon oxide (a-SiOx:H) synthesized via chemically active plasma processing,[258] TiO_2 films,[259,260] magnesium fluorides, and zirconia.[261]

Recently, stimulated by new material properties, a number of pioneering research investigations introduced new concepts to develop semi-transparent, color-tunable, flexible, lightweight, robust solar cells.[262] In semi-transparent solar cells, the efficiency is a compromise with the transparency of the two electrodes and the active layers. For semi-transparent solar cells in building integrated photovoltaics, antireflective, self-cleaning coating are essential to minimize the light loss, thus allowing the correct tuning of the light absorption of the junction and of the electrodes on the basis of acceptable room illumination.

Concerning Si-based solar cells, plasmas are widely used to deposit microcrystalline thick Si layers to fabricate the cell junctions.[263-265] At present, the major fundamental limitation of crystalline Si solar cells is the carrier recombination at the cell surfaces and at the contacts due to the increasing carrier lifetime. To avoid recombination, the Si surface is passivated[266-268] with thin dielectric films. It is possible to carefully tune the properties of the thin passivating layers, thus providing an improved contact where electron- and hole-selective layers largely suppress contact recombination and, at the same time, allow for an effective transport of the majority carriers to the metal contacts.[266]

Plasmas are also utilized to modify the semiconductor surface to tailor the interface states in semiconductor insulator-semiconductor heterojunction solar cells,[269,270] and to change the electrodes characteristics in polymer-based solar cells,[271] as well as the electronic and ionic processes in dye-sensitized solar cells.[272]

Entering deeply in the structure of the solar cell, plasma technology is the key technology to fabricate heterostructures. Beside bulk crystalline silicon-based junctions, Si-based thin films solar cells are of great interest for the substantial reduction of the costs compared to classical crystalline Si solar cells. Essentially, the idea comes from the consideration that the main part of the crystalline

Si material simply acts as a mechanical carrier while the optical absorption occurs in the 30 μm top-layer region. It might then be possible to realize a much thinner crystalline Si film reducing costs while maintaining high electronic quality. This is achieved by means of epitaxial Si film growth on cheap glass or ceramic substrates. Micro- or polycrystalline Si layers are deposited on top of these substrates, and the grain size is determined by the process parameters (growth temperature and supersaturation conditions).

The most widely studied epitaxial deposition technique is the thermally assisted heterogeneous decomposition of an Si and doping precursors at a heated Si surface or thermally assisted chemical vapor deposition (TA-CVD) process[273] and PECVD. This process has been evolved on the basis of the large expertise gained from microelectronics. New epitaxial deposition systems allow highly reproducible depositions in terms of uniform thickness and of dopant concentration. PECVD technology was used to fabricate Si-based heterojunctions composed by a monocrystalline Si substrate passivated by n-type and p-type amorphous hydrogenated Si thin films placed in the front and rear, respectively,[274] or by $SiO_2/a-SiN_x:H$ stacks.[275] PECVD was also utilized to deposit transparent conductive oxide layers providing both collection of carriers and antireflectivity. Conversion efficiency of these kinds of junctions is over 21%.[274] To further increase performances, deposition of thin films becomes crucial for fabricating multijunction solar cells. In ultra-high efficient multijunction solar cells based on III–V elements (In, Ga, As, Ge, Si), efficiencies higher than 40% may be achieved by realizing an ideal combination of bandgaps and lattice-matching as represented in Figure 3.9 for an InGaNAs multijunction solar cell.

The InGaNAs multijunction has a high potential for achieving a narrow bandgap while maintaining lattice matching with Ge or GaAs. However, inhomogeneous distribution of nitrogen (N) leads to a critical drop in carrier mobility. In the past, a homogeneous solid distribution of N into the crystal could be achieved

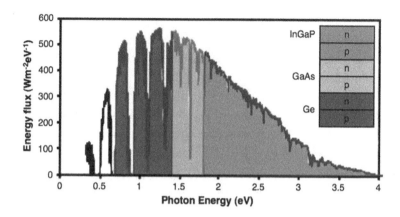

FIGURE 3.9 Solar spectrum radiated on earth and photon flux collected by the top cell (InGaP), middle cell (GaAs), and bottom cell (Ge). (Reprinted with permission from Nakano Y., *AMBIO* (2012), *41* (Supplement 2), 125–131.)

only through molecular beam epitaxy, which is used to fabricate films under high vacuum conditions.[276] Currently, less complex technologies such as PECDV provide thin control over the film thickness of the various kinds of elements which compose the multijunction as well as their doping level.[275-278] In the case of complex structures, plasma etching, atomic layer deposition (ALD) and plasma-enhanced atomic layer deposition (PE-ALD) are utilized to produce the desired stack of layers. Unlike conventional wet chemical deposition methods such as chemical bath deposition, ALD and PE-ALD ensure a thin control of the thickness uniformity all over the substrate surface, which is a crucial parameter for obtaining highly efficient junctions. Furthermore, these techniques are compatible with other plasma deposition techniques such as sputtering and PECVD and the film quality (thickness and uniformity over the absorber surface) is unpaired. In fact, ALD and PE-ALD were utilized for depositing several kinds of oxide,[279-282] sulfides,[283-286] and ternary materials.[287-289] ALD is a gas-phase deposition technique where the precursors, thinly dosed, sequentially react with the substrate surface and then are purged by an inert gas, such as Ar or N_2. Figure 3.10 shows a schematic of the ALD process and the result of the deposition of a thin-film multijunction. Figure 3.10a illustrates the sequential reaction scheme for the ALD growth of ZnO using diethylzinc (DEZ) and H_2O precursors. In Figure 3.10b, a cross-section SEM image of the multilayer structure obtained upon deposition process is shown.

ALD was successfully utilized for the deposition of thin layers composing multijunction solar cells.[290-292] In recent works, it was proven that plasma enhancement is necessary for the synthesis to insert oxygen in indium sulfide films and crucial for the successful implementation of the thin films.[293] Alternative methods to deposit high-quality crystalline layers are the metallorganic chemical vapor deposition (MOCVD) and plasma-enhanced MOCVD. Utilizing a combination of MOCVD and plasma etching, it was possible to fabricate complex multijunction structures as those shown in Figure 3.11.

Finally, plasma processes have been successfully applied to the continuous roll-to-roll production of multijunction Cu(In,Ga)Se$_2$

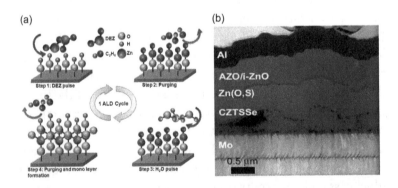

FIGURE 3.10 (a) Schematics of the ALD sequential deposition of thin films; (b) sequence of thin layers deposited using ALD. (Reprinted with permission from Sinha et al., *Solar Ener. Mater. Solar Cells* (2018), *176*, 49–68.)

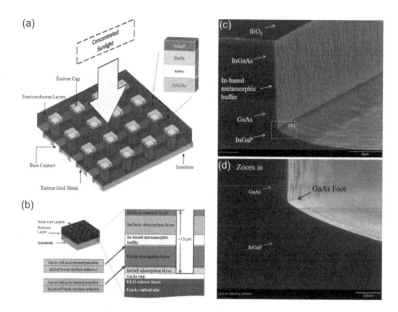

FIGURE 3.11 (a) Schematic diagram of a backside contact InGaP/ GaAs/InGaAs IMM triple-junction solar cell. (Inset) epitaxial hetero- structure. (b) Cross-sectional cartoon of the triple-junction solar cell heterostructure. The overall epitaxial layer thickness for etching the vias is approximately 13 µm. (c) SEM micrographs of the triple-junction solar cell etched with 5 sccm Cl_2 and 10 sccm Ar, and (d) 10 sccm Cl_2 and 20 sccm Ar. (All other process parameters held constant: 100W RIE power, 300W ICP power, 180°C, and 3 mTorr.) (Reprinted with permission from Zhao Y. et al., *J. Vac. Sci. Technol.* (2012), *B30*, 06F401.)

solar cells deposited on flexible substrates.[294] In spite the reference standard efficiency being ~20%, this continuous roll-to-roll process led to an overall efficiency of 10%. Further optimization of the metal composition ratios [Ga]/[Ga] + [In], [Cu]/[In] + [Ga] and the Se con- tent is required to improve the cell performance together with the optimization of the Na passivation process to reduce the presence of interface defects. A detailed analysis of the fabrication process of $Cu(In,Ga)Se_2$ solar cells and the utilization of plasma in the various construction stages may be found in the work by Klinkert.[295]

3.8.2 Application in Energy Production

Pushed by environmental issues, even more frequent is the use of plasma technology for producing energy from waste, biomass, and coal all around the world. Environmental and economic aspects show that the application of this technology is a promising alternative to conventional systems. In particular, plasma can be utilized to increase the efficiency of combustion, gasification, pyrolysis, and reforming processes. A good example is the dry reforming of methane which could be a mean to reduce the anthropogenic impact due to CO_2 emission.[296] Apart from CO_2, sequestration in oil fields or in geological formations, CO_2 can be reutilized to produce more precious molecules such as urea,[297] polycarbonate,[298,299] salicylic acid,[300] cyclic carbonate,[301,302] polypropylene carbonate,[303,304] methanol,[305,306] syngas,[307–309] and other organic compounds. Recently, syngas has received much attention because, being composed of hydrogen and CO, it can be used as fuel in the Fischer–Tropsch process to produce hydrocarbons,[310] such as olefin and paraffin,[311] and liquid fuels.[312]

Among the possible chemical routes utilizing different kinds of catalyzers and temperatures, plasma is another promising technique to synthesize organic molecules. Being formed by charged particles and electrically neutral particles (atoms, molecules, and radicals), it can induce chemical reactions through radical recombination. For this reason, non-thermal plasmas (corona, microwave, spark, discharge, atmospheric-pressure glow discharge, and dielectric barrier discharge) have been investigated to produce syngas. To achieve higher efficiencies in the CO_2 and CH_4 conversion, plasma can be combined with an appropriate catalyzer merging the advantages of catalysis and plasma reforming. In plasma catalysis processes, by inserting a catalyzer in the reactor, it is possible to obtain a double effect: (i) plasma modification of the catalyzer surface, which can induce the catalyst to have higher activity (modification of the catalyzer work function) and better durability; (ii) since catalysts are non-conductive oxides, they influence the charge distribution and the electric field in the plasma. Microdischarges occur in the pores or gaps of the catalysts leading to local high plasma densities, thus enhancing the dissociation and generation of reactive species with beneficial effect on the product generation. This second effect strongly depends on the nature of the catalyzer (its dielectric constant), the reactor geometry, the pressure, and the kind of plasma discharge selected.[296]

Figure 3.12 compares the characteristic timescales in plasma and catalytic reactions. The chemical reactions in plasma can be divided into fast (radicals) and slow (O_3 and H_2O_2) reactions occurring on a timescale of 10^{-6} s and milliseconds respectively.[313] In a catalytic reaction, the initial chemisorption takes place within 10^{-9} s. However, a total adsorption process is needed for the reaction to proceed and this needs much longer times, up to 10^0 s, due to spillover and surface diffusion, along with gas-phase diffusion into the micropores. It is known that the catalytic turnover

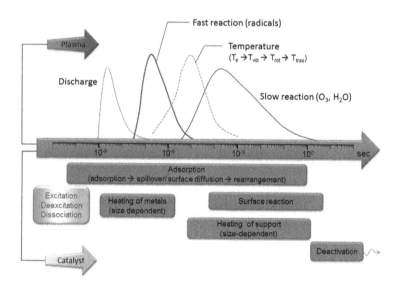

FIGURE 3.12 Timescales in plasma (above) and catalytic chemical reactions (below). (Reproduced with permission from Kim et al., *Plasma Chem. Plasma Process.* (2016), *36* (1), 45–72.)

extends over a timescale of 10^{-2}–10^2 s.[314] This time is 2–3 orders of magnitude slower than the fast radical reactions making the plasma-assisted reactions more efficient.

Thermal plasmas are also utilized for their high energy density, high temperature, and high enthalpy, which makes them useful in a variety of fields such as metallurgy, cutting, welding, etching, and more efficient production of energy from different types of wastes and biomass. The utilization of plasma is attractive especially for the assisted combustion of low-quality matter such as coal, plasma gasification, and pyrolysis,[315,316] or to optimize the combustion[317] also in critical conditions.[318]

Plasma-assisted combustion is also a convenient technology to eliminate all types of wastes including hazardous, industrial, medical, and municipal solid waste.[319–321] In this respect, an interesting application of plasma is for the disposal of plastic waste[322] and waste gasification[323] with energy recovery. The process consists of an extremely high-temperature decomposition of plastics made in an oxygen-starved environment. Avoiding incineration, the process leads to formation of syngas composed by simple molecules such as CO, H_2, and small amounts of higher hydrocarbons. Quenching temperature after decomposition avoids recombination of gaseous molecules with formation of toxic chemical species. Plasma-assisted pyrolysis is utilized to obtain better combustion yields form coal.[324]

In controlled conditions, the presence of decomposition and high temperatures may be exploited to produce desired compounds. As an example, carbon black was produced using an arc discharge non-thermal plasma enhanced by thermal pyrolysis. In this process, propane molecules were cracked into carbon black and hydrogen. Compared to conventional high-temperature processes (1700 K–2500 K), this synthesis occurs at much lower temperatures, thus leading to a significant reduction of the energy consumption. In addition, the conversion efficiency in plasma-assisted pyrolysis is much higher, limiting the raw material waste due to incomplete combustion and emission of environmentally dangerous gaseous molecules such as CO_2, NO_x, SO_x. Finally, plasma carbon black can be produced also from the decomposition of hydrocarbon (methane, ethane, and propane), thus overcoming the need of highly aromatic molecules needed by the classical processing.[325] In another process, acetylene was produced from the pyrolysis of pulverized coal in a rotating plasma reactor,[326] while plasma reforming of methanol allowed for the production of hydrogen for on-board applications.[327] Recently, plasma was utilized in an advanced plant in which waste-to-energy plants and electric energy storage systems are combined.[328] The system is formed by an electric energy storage section based on a solid oxide electrolyzer, a plasma-based technology for waste gasification, and a solid oxide fuel cell for power generation. System efficiency in terms of electricity production, waste-to-energy conversion, and electric energy storage of renewable sources was evaluated by using numerical models. Results highlighted high electric efficiency (35–45%) and, thanks to the use of plasma which is crucial for waste-to-energy conversion, the storage efficiencies are very impressive (in the range of 72–92%).

3.9 COMBINATORIAL MATERIAL PROCESSING BY PLASMAS

Computational modeling or material rational designing on *a priori* information to optimize material properties are very attractive because, in principle, they could guide the material synthesis avoiding production and characterization of numerous unsuitable candidates. However, *ab initio* and other theoretical modeling require even more complicated and time-consuming computations due to the increasing complexity of the materials. In the case of rational design, they need detailed knowledge of the relations between materials' properties and a set of correspondent performances. However, the increasing complexity of the materials leads to an inaccurate description of their properties even if based on a large number of experiments and simulations.[329] As observed by Potyrailo et al., "Thus, in addition to limited examples of rational materials design, a variety of materials have been discovered using detailed experimental observations or simply by chance, reflecting a general situation in materials design that is still too dependent on serendipity."[330] The failure of computational and rational design methods stems from the consideration that, in addition to composition, the materials properties depend also on a series of physical parameters related to the material-preparation conditions. A combinatorial and high-throughput (CHT) approach to material synthesis was then proposed as an alternative method to solve these problems, allowing the production of new materials and significant progress in knowledge.[331–341]

The first concept of CHT approach to material synthesis was first proposed by Hanak.[342] The concepts proposed in this article were truly ahead of their time, and only 25 years later, those ideas were realized.[343] A typical combinatorial materials development cycle is outlined in Figure 3.13.

Compared to the original concept of combinatorial process (see Figure 3.13a), the modern workflow (Figure 3.13b) shows several new important aspects. The modern CHT method is a process constituted by (1) a design/planning of experiments, (2) the creation of sample libraries, (3) high-throughput characterization, and (4) data mining to model the results in the form of structure–property relationships.

The power of the CHT approach is the possibility to prepare sample "libraries" containing hundreds to thousands of variable combinations, thus deeply exploring different combinations of physical/chemical parameters. Testing large material libraries requires high-throughput characterization methods of the chemical and physical properties. Different high-throughput characterization tools are then required for rapid and parallel assessment of single or multiple properties of the large number of samples fabricated together in a CHT process.[344,345] As a result, in a reliable combinatorial workflow, the material performances should correlate well with those obtained utilizing traditional synthesis and testing methods. The benefits of CHT methods include (i) more

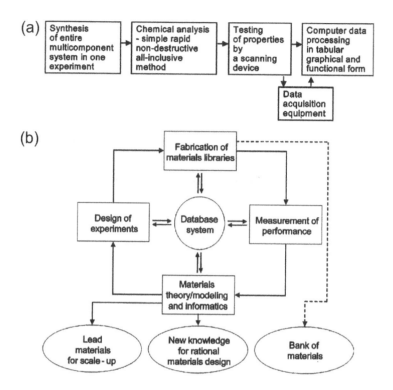

FIGURE 3.13   (a) The original combinatorial approach as proposed by Hanak[342]; (b) the modern scheme of high-throughput combinatorial process. (Reproduced with permission from R. Potyrailo et al., *ACS Comb. Sci.* (2011), *13*, 579.)

efficacious characterization of thermodynamic and kinetic behavior, (ii) more efficient identification of structure–property correlations, (iii) faster development of functional materials, and (iv) reduced experimental variance due to synthesis in the same environmental conditions.

The first attempts to use the CHT approach to synthesize inorganic materials was made by Kennedy et al. The authors describe the development of novel ternary alloys.[346] The CHT was applied to determine isothermal sections in ternary-alloy diagrams. Fe, Cr, and Ni were evaporated simultaneously onto a heated substrate to produce the alloy with simultaneous variation of the alloy composition from point to point. This allowed the authors to rapidly obtain the phase diagram of the ternary compound.

An undoubted area where the CHT approach may speed up the introduction of new materials is electronics. The continuous requirement of higher-performance electronic circuits calls for the use of new materials enabling the reduction of the single electronic component dimensions and thus higher integration levels. An example of the application of the CHT to the development of high-k dielectrics is an integrated circuit transistor device where the dielectric barrier is utilized to separate the gate from the source and the drain. *High-k* gate dielectric materials have been synthesized mixing HfO_2-Y_2O_3-Al_2O_3 in different concentrations.[347–349] Figure 3.14 shows an example of the deposition obtained by using a pulsed KrF excimer laser at a power of 3 J/cm² to ablate the Al_2O_3, HfO_2, and Y_2O_3 targets at a temperature of 300°C. 100 nm-thick films were obtained with variable composition.

The dielectric properties of the deposited film were obtained using a microwave microscope and X-ray diffraction. By adjusting the laser spot size and fluence, the ablation and consequently the composition and thickness of such films can be tuned. PLD was also used to produce a combinatorial library of high-k dielectric films based on the ternary combination of HfO_2–TiO_2–Y_2O_3.[350] The problem is to synthesize a material with a dielectric constant higher than that of HfO_2, which is 20. TiO_2 has a higher dielectric constant, $k \sim 80$, but it is unstable on Si.[351]

Varying the deposition conditions of HfO_2, TiO_2, and Y_2O_3, the authors were able to obtain not only control over the material amorphous and crystalline phases (see Figure 3.14d) but also a material possessing the required thermal stability on Si. The dielectric constant k relative to the composition library ranged from 10 up to 120, which surprisingly was observed near the TiO_2–Y_2O_3 binary system.[350]

High-k materials were also produced by using CVD processes and with ALD.[352] Combinatorial libraries were obtained

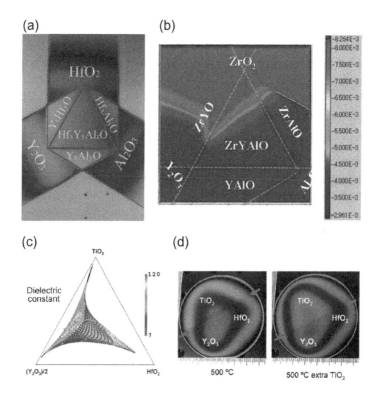

FIGURE 3.14 (a) HfO_2–Y_2O_3–Al_2O_3 ternary composition spread film grown on a Si(100) substrate. The central triangle corresponds to the ternary phase diagram while the three isosceles regions around it correspond to three binary-phase diagrams; (b) dielectric property mapping obtained by measuring the frequency shift on the ternary composition with a microwave microscope. (a,b: Reproduced with permission from K. Hasegawa et al., *Appl. Surf. Sci.* (2004), 223, 229.) (c) Ternary diagram showing the predicted compositional coverage (small points) of the 500°C HfO_2–TiO_2–Y_2O_3 library film and the measured relative dielectric constant (filled points with different gray intensities) extracted from the current voltage characteristics of the as-deposited film. (d) Photographs of HfO_2–TiO_2–Y_2O_3 combinatorial depositions, left at 500°C, right at 500°C with 50% more TiO_2 deposited. The different position of the boundary separating the TiO_2-rich region from the HfO_2–Y_2O_3 is indicated by the arrows. (Reprinted with permission from Klamo J.L. et al., *J. Appl. Phys.* (2010), *107* (5), 054101.)

varying the gaseous precursor flow during the plasma process.[353]

CHT was also applied to generate new metal gate electrodes. Magnetron sputtering and masks were used to deposit Ni–Ti–Pt ternary compounds on HfO_2 dielectrics[354] leading to controlled changes of the material work function. Similar experiments were carried out using Mo/Ru and Pt/W combinations,[353,355] as well as metal nitrides and carbides.[356-358] Another fervent area of activity is the development of magnetic materials useful for high-temperature superconductivity, colossal magnetoresistance, spin polarization, and long spin relaxation times useful for spintronics and other kinds of applications. Shadowing masks were utilized to synthesize binary and ternary epitaxial combinatorial libraries based on lanthanides.[359,360] The structural-magnetic-electronic properties of the material libraries are then investigated via CHT as a function of component concentration with XRD and scanning SQUID probes at low temperature, providing structural maps and magnetic domain distributions.[359] In other works, magnetooptic imaging was applied for mapping the colossal magnetoresistance diagram for the $La_{1-x}Ca_xMnO_3$ films deposited on a $SrTiO_3$ substrates[361] and on $(Zn,Co)Fe_2O_4$ materials. The power of the CHT was also demonstrated with the discovery of magnetic semiconducting oxides in $Co-TiO_2$.[362]

Magnetic alloys were produced using co-sputtering techniques and rapidly characterized by phase mapping. Materials such as $(Ni_xFe_y)_{1-x}(SiO_2)_x$,[363] Fe-Ni-Co,[364] Ni_2MnGa, Co_2MnGe,[365,366] and Ni-Mn-Ni_2 Ga_3 were investigated by synthesizing material libraries. As an example, Figure 3.15 illustrates a mapping of magnetic properties using a room-temperature scanning SQUID microscope.[367]

The discovery of high-temperature superconductivity in the mid-1980s stimulated intense research activities. CHT has demonstrated to be effective also in this area. Libraries of binary thin films have been developed starting from combinations of $BaCO_3$, Bi_2O_3, CaO, CuO, PbO, $SrCO_3$, and Y_2O_3 with different stoichiometries,

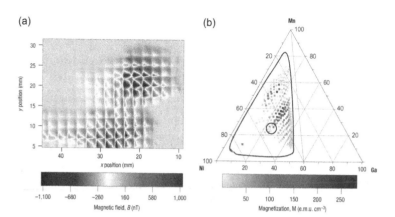

FIGURE 3.15 (a) Scanning SQUID microscope image of 2×2 mm array of a Ni–Mn–Ni$_2$Ga$_3$ thin film deposited on a Si wafer. The microscope is sensitive to magnetic poles which are more intense in the highly contrasted zones. (b) Room-temperature magnetic-phase diagram of Ni–Mn–Ga. The region inside the black curve maps is the composition of the deposited film. The composition where the central straight line (from Mn vertex to 60) meets the Ni–Ga line at 60, is Ni$_2$Ga$_3$ (one of the three target compositions used). The circle marks the compositions near the Ni$_2$MnGa Heusler composition. (Reproduced with permission from Takeuchi I. et al., *Nat. Mater.* (2003), *2* (3), 180.)

and were generated with a series of binary masks.[343] Deposition of samples of 200×200 μm size of cuprate were characterized using high-resolution scanning susceptibility and Eddy current detectors, allowing the identification of superconductivity in BiSrCaCuO and YBaCuO films. In the studies of Hanak and Jin et al., the combinatorial approach was used to explore the super-conductive properties of the materials.[368,369] Also in these cases, the degree of the superconductivity mirrored by the change of the material resistivity with temperature was associated to the change of material composition. Then, for rapid screening of new super-conductors, the parallel four-terminal resistance-temperature measurement method can be utilized.[370]

With respect to electronic properties of materials, the optical ones are among the most important. Lighting and imaging are based on phosphorescent materials, as well as plasma, field emission, and electroluminescent displays. The utility of the CHT was first demonstrated by Sun et al. The authors used an RF sputtering apparatus

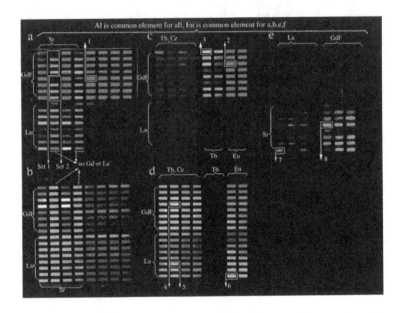

FIGURE 3.16 Photoluminescent image of a series of phosphor libraries: different shades of gray indicates different colors and different emission intensities. Indicated are the correspondent nominal compositions obtained via different deposition conditions: (a) La (or $GdF_3)_m(Sr)_n AlO_x:Eu_y^{2+,3+}$ where $0.375 \leq m \leq 1$, $0.25 \leq n \leq 0.4$, $1.88 \leq y \leq 12\%$ in atomic ratio, were annealed at 1150°C in 10% H_2/Ar for 4 h; (b) same as (a), but annealed at 1400°C in 40% H_2/He for 4 h; (c) La (or $GdF_3)_m AlO_x:Tb_y^{3+}$ (Ce_z^{3+}) $Eu_h^{2+,3+}$ where $0.32 \leq m \leq 1$, $1.29 \leq y \leq 6\%$, $0.65 \leq z \leq 4\%$, $1.29 \leq h \leq 8\%$ in atomic ratio, were annealed in air at 1150°C for 4 h; (d) same as (c) but annealed at 1400°C in 40% H_2/He for 4 h; (e) La (or $GdF_3)_m(Sr)_n AlO_x:Eu_y^{2+,3+}$ where $0.178 \leq m \leq 0.714$, $0.17 \leq n \leq 0.4$, $0.75\% \leq y \leq 16.7\%$ in atomic ratio, were annealed at 1150°C in 4% H_2/Ar for 4 h. (Reprinted with permission from Sun X.-D. et al., *Appl. Phys. Lett.* (1997), *70* (25), 3353.)

and a set of physical masks to synthesize thin-film libraries based on Gd(La,Sr)AlOx metal oxides.[371] Figure 3.16 shows the photograph of the material libraries taken with a UV lamp. Indicated are also the relative elemental compositions and stoichiometries.

In another work, 1024 different compositions were produced to generate a library of silicate and gallate host materials containing a dopant selected among CeO_2, EuF_3, Tb_4O_7, Ag, TiO_2, and Mn_3O_4.[371] The combinatorial library was obtained through a masking strategy and photolithography, leading to compositionally diverse thin-film libraries on 2.5 cm^2 substrates. Optimal compositions were identified with the use of a parallel investigation of the photoluminescence properties of the single elements. Color variations arise from different material composition, different thicknesses, and different deposition conditions (different deposition temperatures).

Another interesting area where CHT applies is the discovery of new transparent conducting oxides (TCO). Diverse are the applications of TCO which include photovoltaic (PV) cells, touch screens, flat-panel displays, and light-emitting diodes. In photovoltaics, TCO is used as a front electrode. Then essential properties of a TCO are both the low resistivity which should be $<5 \times 10^{-4}$ Ω-cm (or a conductivity ≥ 2000 (Ω-cm)$^{-1}$) and transparency in the solar spectrum range (>80% from 300 nm to 1200 nm wavelengths). Today TCO are generally based on crystalline In_2O_3 doped with Sn with Sn/In = 2% which, at room temperature, has a resistivity of 2×10^{-4} Ω-cm, and the electron concentration is $3 \times 10^{20}/cm^3$.[372] Great interest is addressed to the synthesis of amorphous In based alloys because the simplified and cheaper production processes. With this aim, extended research has been devoted to alloying In_2O_3 with zinc (Zn) and Sn to form amorphous zinc indium oxide (ZIO) and tin indium oxide (TIO). Since the amorphous material structure leads to lower conductivities than the crystalline ones, CHT has been used to optimize the material properties and obtain a compositional phase diagram.[373,374] The transparency of the synthesized materials was tested utilizing light sources at wavelengths from UV to NIR while Hall measurements were performed to test the conductivity. ZIO materials showed the lower resistivity for Zn concentration in a range from 45% to 80%.[375] Significant improvements in the conductivity were obtained doping the In_2O_3 with molybdenum, titanium, zirconium, and Ti. In the first case, the $Mo_x:(In_2O_3)_y$ material libraries led to a maximum conductivity of ~1000 $(\Omega\text{-cm})^{-1}$ at an Mo concentration of 6 at%.[376] This limited value was ascribed to the opposing trends of the carrier concentration and the electron mobility. Much better results were obtained in Ti-O-In materials, which, for optimized concentrations ranging from 3 at% to 4 at% and after annealing at 500°C, led to a conductivity of ~6000 $(\Omega\text{-cm})^{-1}$.[374] Using combinatorial PLD depositions of $In_{2-2x}Me_{2x}O_3$ where Me = Ti, Zr, or Sn, it was found that better material transparency is obtained with Ti and Zr doping without affecting the conductivity.[377] Other material libraries based on SnO_2 or ZnO were produced using CHT. SnO_2 is generally doped with fluorine or antimony. In particular, $F:SnO_2$ CHT tests were carried out using an atmospheric pressure chemical vapor deposition instrument. The apparatus allowed the treatment of large Si wafer and local heating, thus enabling the deposition of a matrix of films at different temperatures and precursor mixtures. In the best deposition conditions, the TCO showed a transparency of 99% with acceptable levels of resistance.[378] In another work, using a CVD combinatorial process, authors were able to deposit $F:SnO_2$ TCO library, changing the dopant concentration and the substrate temperature during the deposition.[379] Optimal TCO properties were found for a substrate temperature of 500°C and a F concentration of 2.2 at.%. In these deposition conditions, the sheet resistance was ~3 (Ωsq) with a charge carrier density of $(6.4 \times 10^{20}$ $cm^{-3})$. Thanks to the pyramidal texture of the TCO film surface, transparency was always about 80% from UV to NIR. ZnO is another commonly used oxide to synthesize TCO. ZnO displays high chemical and thermal stability while providing conductivities as high as 5×10^3 $(\Omega\text{-cm})^{-1}$, and much cheaper and less toxic than indium. Different dopants such as gallium, aluminum, nitrogen, and cobalt were utilized in combination with ZnO to produce TCO.[380–384] Figure 3.17 shows a combinatorial deposition of gallium, doped ZnO obtained using a continuous composition spread method.[380] The high-throughput approach allowed the identification of optimized Ga:ZnO composition leading to the deposition of thin films with resistivity as low as 1.46×10^{-3} (Ω-cm) and an average transmittance above 90% in the 550 nm wavelength region. In particular, Figure 3.17a correlates the deposition position with the correspondent transmittance locally measured utilizing an optical fiber. For six sampled points of the TCO, Figure 3.17b shows how changing the radiation energy changes the absorption coefficient α, $\alpha h\nu = B(h\nu - E_g)^{1/2}$ where E_g is the optical gap of the film, $h\nu$ the photon energy, and B the constant relating to the electron–hole mobility.

CHT methods were utilized also to develop materials for energy applications, which will described very synthetically here.

The CHT approach was applied to produce anodes and cathodes for lithium ion batteries. Due to their superior gravimetric and volumetric energy densities, and cycling properties, batteries based on Li have been identified as the actual best power source for portable electronics, laptops, cell phones, and high-performance electric vehicles. Actual Li ion batteries are characterized by 2400 mA-h energy storage capacities, 250 W-h/kg specific energies; 620 W-h/l energy densities, and a specific power of 340 W/kg. In Li batteries, the anode is composed by graphite which intercalates Li ions provided by $LiCoO_2$ anodes during discharge. Combinatorial synthesis was applied to discover novel materials suitable for high-performance anode and cathodes. As for cathodes, promising materials are based on Li-Ni-Co ternary oxide or Li-Ni-Co-Ti-O and other $LiMe_xO_y$ compounds, where Me = Al, Co, Cr, Mn, Ni, Ti.[385–390] Concerning anodes, great effort was devoted to introduce Sn in the anodes to improve their resistance. Sn-C, Sn-Co, Sn-Ti, and Sn-V amorphous materials have been explored.[391,392] CHT investigations were also performed on anodes composed by Si-Al-Mn alloys.[393,394] However, difficulties were found in measuring the material

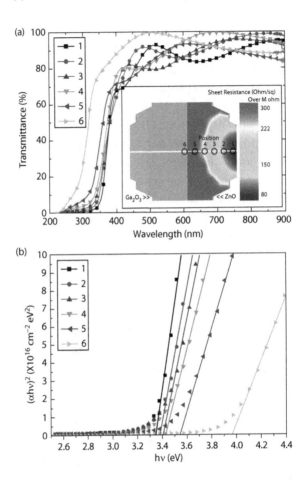

FIGURE 3.17 (a) The optical transmittance and (b) the $(\alpha h\nu)^2$ versus $h\nu$ plot measured in the six positions of Ga:ZO thin films, indicated by circles in panel (a). (Reprinted with permission from Jung K. et al., *Appl. Surf. Sci.* 2010, *256* (21), 6219.)

conductivity when the dimensions of the single testing electrode were reduced to increase the number of parallel tests.

PLD and continuous composition spread were also utilized to develop thermoelectric materials. Combinatorial material libraries were obtained using Ca-Ba-Co oxide,[395] Ca-Sr-La-Co$_4$O$_9$,[396] Ce-Co-Sn,[397] Mg-Si-Ge ternary systems,[398] and more traditional

Ni-Cu binary alloys.[399] Optimization of the conversion efficiencies was performed applying the CHT approach, aimed at maximizing the figure of merit, ZT, which is equal to $S^2\sigma T/k$ where S is the Seebeck coefficient, σ is the electrical conductivity, k is the thermal conductivity, and T is the absolute temperature. For the various depositions mentioned, the conductivity and Seebeck coefficients are: Ca-Ba-Co: $\sigma = 30$ S/cm and $S = 120$ µV/K; Ce-Co-Sn: $\sigma = 10^5$ $(\Omega cm)^{-1}$ and $S = -15$ µV/K; Ca-Sr-La-Co_4O_9: $\sigma \sim 60$ $(\Omega cm)^{-1}$ and $S = 75$ µV/K; Mg-Si-Ge: $\sigma \sim 500$ $(\Omega cm)^{-1}$ and $S = -45$ µV/K.

An important area to be explored with a CHT approach is the synthesis of materials for hydrogen storage. Energy storage is considered one of the key and most challenging objectives for achieving an economy based on renewable energy sources. At present, there are no efficient methods to store large amounts of electricity. For this reason, one promising way is to accumulate energy in chemical form using hydrogen. Hydrogen (H_2) can be stored at high pressure or liquefied, but both of these solutions pose some important issues related to safety and energy employed for compression or liquefaction. The chemical route appears then as a promising opportunity to store hydrogen in significant amounts. In this respect, H_2 can form hydrides with Li, B, Mg, N, Al, and Na leading, for some of them, to storage densities as high as 18 wt.%.[400] Unfortunately, the materials showing high storage H capacity such as $LiBH_4$ are accompanied by high working temperatures (650°C for $LiBH_4$),[401] which are detrimental for the energy balance and present some technological problems for applications in automotive and in portable devices. CHT technology can then be of great help in the search of suitable efficient H storage materials. In this respect, the introduction of transition metals was utilized to improve the kinetic performances of the materials.[401] Essentially, transition metal catalysts speed up the dissociation of hydrogen molecules, thus improving the H sorption kinetics.[402] Pd is one of the best catalyzers for H_2 molecule dissociation. Placing Pd clusters strongly improves the hydrogenation/de-hydrogenation kinetic of Mg[402] or more complex Mg_2Ni,

LaNi$_5$, FeTi systems.[403] However, the high cost of Pd prevents industrial applications. A more convenient catalyst is Ni, which shows high affinity for hydrogen. The introduction of 1 at.% of Ni in Mg leads to a 50% increase of the storage capacity and a reduction of the H$_2$ desorption, adsorption temperatures from 275°C to 175°C and from 350°C to 275°C respectively.[404] Besides Ni, Ti, and V and their combination, Iwakura et al. were able to increase the discharge capacity of Mg$_2$Ni-Ni composite over multiple cycles by increasing the fraction of the material amorphous phase.[405] Thus it makes sense to explore materials and catalysts such as a mixture of LiBH$_4$:MgH$_2$,[406] LiNH$_2$: MgH$_2$,[407] and more complex alloys,[408,409] using the combinatorial approach. This would be impossible using the conventional material science techniques because of the high number of chemical, physical, and thermodynamic parameters that can be changed during the material synthesis. Mg-Ni binary,[410] Mg-Ni-Fe, and ternary-alloy libraries[411] were synthesized and hydrogenation investigated through IR emissivity maps.[410] XRD[412] and pressure-optical transmission isotherm[413,414] (see Figure 3.18) mapping can also be utilized to investigate the change in material structure upon formation of the hydride.

Hydrogenography was also applied to characterize other Mg-based binary and ternary alloys such as Mg-Al,[415] Mg-Al-Ti,[416] and Mg-Ni-Ti.[417] In this last case, the CHT approach led to the identification of the best composition: Mg$_{69}$Ni$_{26}$Ti$_5$ possessing an hydrogenation enthalpy of ~39.2 kJ/mol H$_2$, very similar to the value of ~40 kJ/mol H$_2$ ideal for on-board hydrogen storage in vehicles.[413] Another interesting method for combinatorial screening of hydrides is based on the measurement of the expansion/contraction of the material during cycling using of microelectromechanical cantilevers.[418,419] An example is shown in Figure 3.19.

Another area where CHT was successfully applied is relative to the optimization of biomaterials for specific applications. The synthesis of a biomaterial involves the possibility to tune the surface roughness as well as the surface chemistry, both of which play

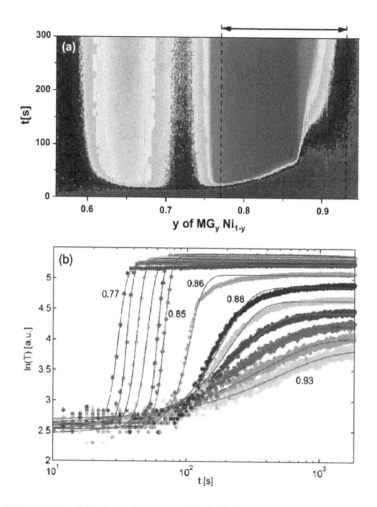

FIGURE 3.18 (a) False color map of ln(T), the logarithm of the transmission, as a function of time and composition y of $Mg_y Ni_{1-y}$ after the application of a $p(H_2) = 2020$ Pa pressure step at temperature T = 333 K (first loading). Dark gray lateral regions represent low transmission while mid-gray central regions represent high transmission. (b) Symbols ln(T) as a function of time for various compositions between $0.77 < y < 0.93$ (the values indicated on the figure correspond to the arrow in (a)). The full lines are Boltzmann fits used to determine the hydrogen absorption rate. (Reproduced with permission from Gremaud R. et al., *Acta Mater.* (2010), *58*, 658.)

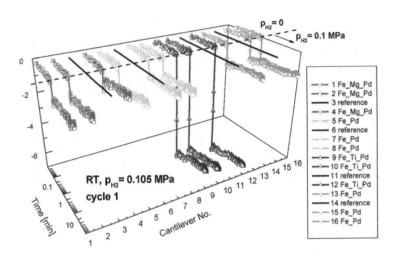

FIGURE 3.19 Hydrogenation level simultaneously measured on 16 thin-film/cantilever combinations (Si/SiO$_2$/Si$_3$N$_3$–5 nm Fe–x nm metal–10 nm Pd) at RT; 12 cantilevers are coated as indicated, 4 uncoated as reference. (Reprinted with permission from Ludwig A. et al., *J. Alloys Compd.* (2007), 446, 516.)

a central role in the process of cell adhesion. Biomaterials relate to a class of materials including biopolymers, ceramics, metals, and composites of organic, organometallic, and inorganic compounds[420,421] that are used to fully or partially replace organs and to make prosthesis and ligaments/tendons. They are also utilized for heart valves and blood vessels, drug delivery, artificial skin, bone cements, and dental implants.[421] Severe requirements have been set in place by the International Organization for Standardization (ISO) in relation to material biocompatibility.[422] Standardized tests have to be performed to determine the degree of biocompatibility/toxicity of a biomaterial and the CHT approach may substantially facilitate the evaluation of the multiple parameters governing the biomaterial/living tissue interplay.[423–425] Ideally, a scaffold should interact with cells thus regulating the complex processes of tissue formation and regeneration. In particular, when any biomaterials come in contact with a biological environment, a rapid adsorption of proteins on the substrate surface takes place.[426] Then the membrane integrin receptors start to interact with the extracellular matrix proteins present on the substrate.[427] The kind of protein and its amount and conformation, are responsible for cellular adhesion and proliferation, as well as for the cell differentiation

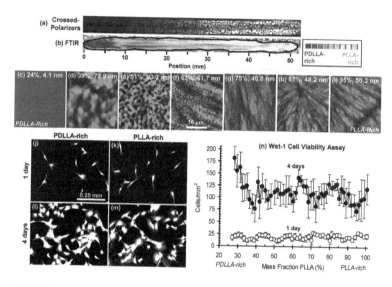

FIGURE 3.20 PDLLA/PLLA strip gradient libraries: cell screening. (a) An image of the birefringence from a PLLA/PDLLA gradient is shown. The PLLA-rich portions possess higher crystallinity and more birefringence than the PDLLA-rich areas. (b) Correspondent Fourier transform infrared (FTIR) map of the PDLLA/PLLA gradient library. Different gray tones are used to identify PDLLA-rich regions (dark gray) PLLA-rich regions (light gray). (c–i) AFM topographs from different positions in the PDLLA/PLLA gradients showing different surface roughnesses. The scale bar in (f) applies to all topographs. The height range is 500 nm; (j–m) Fluorescence microscopy images of MC3T3-E1 preosteoblasts cultured for 1 day (j, k) or 4 days (l, m) show that cell morphology is not influenced by PDLLA/PLLA gradients. (n) The cell number on the gradients: adhesion at one day was unaffected by composition while proliferation at 4 days was enhanced on the PDLLA-rich regions. Error bars are S.D. (n = 6). (Reproduced from Simon J.C. et al., *Biomaterials* (2005), *26*, 6906.)

and possible migration[428] (see Figure 3.20). The CHT approach can be of great help allowing modulation of the substrate surface properties during deposition. This enables the possibility to produce process libraries where materials with surface chemistry and the surface topography can be thinly changed.

As an example, a material library made by combining poly(L-lactic acid) (PLLA) and poly(D, L-lactic acid) (PDLLA),[429] poly(ε-caprolactone) (PCL) and PDLLA[424] or poly(DTE carbonate) and poly(DTO carbonate)[425] was utilized to make composition gradients, leading to different surface porosity. The authors found that the cell attachment depends on both the roughness and the composition of the substrate, likely suggesting that they influence both the amount and the conformation of proteins triggering the cell adhesion.[430,431]

Combinatorial synthesis was also applied to discover new material formulations for regenerative medicine, in particular to trigger the growth of stem cells for tissue regeneration and repair. For example, hydrogels were tested to determine the best material properties to promote the myofibrogenesis of mesenchymal stromal for heart valve tissue regeneration.[432] Soft, more porous, and readily degradable hydrogels were identified as the best to trigger myofibroblastic cells growth expressing high levels of α-smooth muscle actin and collagen type I. Similarly, high-throughput screening of inorganic biomaterials (SiO_2, Ti/TiO_2, Cr/CrO_3, Al_2O_3) varying the surface topography was applied to determine optimal conditions for human bone marrow-derived mesenchymal stem cell growth and proliferation.[433] This study found the optimal wrinkle dimension (wavelength: 7121 nm; amplitude: 2561 nm) for promoting cell alignment and focal adhesion assembly and orientation, and for the development of long/parallel filopodia.

In another study, the CHT method was applied to evaluate the effect of combinations of extracellular matrix proteins (collagen I, collagen III, collagen IV, laminin, fibronectin, and elastin) on the adhesion and spreading of mesenchymal stem cells.[434,435]

In other *in vitro* work, CHT was utilized to study the effect of multiple growth factors (BMP2, BMP6, GDF5, TGF-β1, and FGF2) on chondrogenic differentiation of human mesenchymal stem cells.[436] This approach has recently developed into methods involving laboratory automation, high-throughput synthesis, and characterization of small molecules libraries to facilitate the synthesis of materials and molecules.[437,438] In this respect, the combinatorial approach was utilized to efficiently design biomaterials for gene,[439] drug,[440] and retroviral drugs[441] delivery. The high-throughput approach enabled more efficient delivery of drugs or of combinations of drugs, leading to superior treatment potency.

CHAPTER 4

Concluding Remarks

THE SURFACE IS THE place where a material interacts with the external environment. Then, modifying the surface properties strongly influences the behavior of that material and how it reacts to the external chemical and physical stimuli. Because plasma treatments are applied to modify the surface properties, this technology has become an essential part of material science. Plasma-assisted functionalization and film depositions have become established processes in a variety of applications.

They are utilized to clean surfaces, remove contaminants, and enhance adhesion by increasing the surface wettability. Grafting of chemically active functional groups such as amine, carbonyl, hydroxyl, and carboxyl groups are used to improve bondability on substrates such as glass, polymers, ceramics, and various metals. Cleaning is also utilized to remove the bacterial load, thus obtaining a more efficient surface sterilization with respect to conventional methods. In fact, plasma technology is applied not only in the sterilization of medical tools but also in the food and packaging industry to avoid product degradation. High density of radicals and strong electric fields lead to complete eradication of bacterial colonies.

Deposition of hard coatings is another important and consolidated application of the plasma technology to preserve materials from consumption. These are applied to protect soft substrates such as polymers or mechanical parts subjected to high friction and hard-working conditions characterized by high forces applied, high contact pressures, and high temperatures, or for protection from intensive chemical attack at the contact zone in machining processes of difficult-to-cut materials. Hard coatings are also used as anti-scratch protective coatings. These are also deposited by plasmas to passivate surfaces, which render them chemically inert and resistant against corrosion.

In the area of fashion and technical textiles, plasma processing is utilized to facilitate the fiber dyeing and, at the same time, to make stain-resistant fabrics.

Plasma treatment techniques are widely utilized in biomaterials engineering. In addition to increasing the wear resistance of orthopedic joints, plasma treatments are used to increase the osseointegration of prosthetic and orthodontic devices. With regard to polymeric biomaterials, the process selectively modifies the surface energy, increasing or reducing the surface wettability. Plasma treatments are applied to increase the material biocompatibility to better mimic the local tissue environment without altering the bulk attributes. In general, plasma treatments provide surface activation, stimulating cell proliferation and differentiation, which is essential in regenerative medicine.

Due to the increasing need for reducing emission and environmental pollution causing global warming, plasma technology has a considerable scientific, technological, and economic–environmental impact. In particular, the high control on the composition and on the thickness of the deposited films makes the plasma one of the most important technologies to produce solar cells.

Reduction of pollution is also obtained by applying plasmas to increase the efficiency of combustion or reducing emissions with energy recovery. The use of plasmas for waste management and reforming processes is an interesting development.

Finally, plasmas are successfully applied to the combinatorial synthesis of materials that accelerates material discovery with more efficient identification of structure–property correlations, and faster development of functional materials prototyping and transfer to industrial production.

As it appears from this review, plasmas are widely applied either for fundamental scientific research and for the manufacture of everyday life products.

In conclusion, plasmas constitute a technology that can revolutionize the material industrial processing, opening unpredictable developments. The possibility of transferring the results from scientific research to industrial applications makes plasma appealing and an evergreen vital technology.

References

1. MacDonald, A. D. *Microwave Breakdown in Gases*; John Wiley & Sons, New York, NY, USA, 1966.
2. Kortshagen, U. Nonthermal Plasma Synthesis of Semiconductor Nanocrystals. *J. Phys. D* 2009, *42* (11), 113001.
3. Chen, X.; Cheng, Y.; Li, T.; Cheng, Y. Characteristics and Applications of Plasma Assisted Chemical Processes and Reactors. *Curr. Opin. Chem. Eng.* 2017, *17*, 68–77. https://doi.org/10.1016/j.coche.2017.07.001.
4. Thiry, D.; Konstantinidis, S.; Cornil, J.; Snyders, R. Plasma Diagnostics for the Low-Pressure Plasma Polymerization Process: A Critical Review. *Thin Solid Films* 2016, *606*, 19–44. https://doi.org/10.1016/j.tsf.2016.02.058.
5. Robertson, J. Thermodynamic Model of Nucleation and Growth of Plasma Deposited Microcrystalline Silicon. *J. Appl. Phys.* 2003, *93* (1), 731–735. https://doi.org/10.1063/1.1529090.
6. Kong, H. CVD Diamond Films: Nucleation and Growth. *Mater. Sci. Eng.* 1999, 32.
7. von Keudell, A. Surface Processes during Thin-Film Growth. *Plasma Sources Sci. Technol.* 2000, *9* (4), 455–467. https://doi.org/10.1088/0963-0252/9/4/302.
8. Girshick, S. L. Particle Nucleation and Growth in Thermal Plasmas. *Plasma Sources Sci. Technol.* 1994, *3* (3), 388–394. https://doi.org/10.1088/0963-0252/3/3/023.
9. Boufendi, L.; Bouchoule, A. Particle Nucleation and Growth in a Low-Pressure Argon-Silane Discharge. *Plasma Sources Sci. Technol.* 1994, *3* (3), 262–267. https://doi.org/10.1088/0963-0252/3/3/004.
10. Bogaerts, A.; Eckert, M.; Mao, M.; Neyts, E. Computer Modelling of the Plasma Chemistry and Plasma-Based Growth Mechanisms for Nanostructured Materials. *J. Phys. Appl. Phys.* 2011, *44* (17), 174030. https://doi.org/10.1088/0022-3727/44/17/174030.

11. Seshan, K. Ed. Intel Corporation Santa Clara, California. *Handbook of Thin-Film Deposition Processes and Techniques Principles, Methods, Equipment and Applications*, 2nd Edition; William Andrew Publishing, Norwich, NY, USA, 2002.
12. Fridman, A. *Plasma Chemistry*; Cambridge University Press, New York, USA, 2008.
13. Chen, X.; Cheng, Y.; Li, T.; Cheng, Y. Characteristics and Applications of Plasma Assisted Chemical Processes and Reactors. *Curr. Opin. Chem. Eng.* 2017, *17*, 68–77. https://doi.org/10.1016/j.coche.2017.07.001.
14. Bonizzoni, G.; Vassallo, E. Plasma Physics and Technology; Industrial Applications. *Vacuum* 2002, *64* (3–4), 327–336. https://doi.org/10.1016/S0042-207X(01)00341-4.
15. Bárdos, L.; Baránková, H. Cold Atmospheric Plasma: Sources, Processes, and Applications. *Thin Solid Films* 2010, *518* (23), 6705–6713. https://doi.org/10.1016/j.tsf.2010.07.044.
16. Chapman, B. *Glow Discharge Processes: Sputtering and Plasma Etching*; Wiley, New York, USA, 1980.
17. Chen, F., Chang, J.P. *Lecture Notes on Principles of Plasma Processing*; Springer, New York, USA, 2003.
18. Chu, P. K.; Lu, X. *Low Temperature Plasma Technology: Methods and Applications*; CRC Press, Boca Raton, FL, 2014.
19. Conrads, H.; Schmidt, M. Plasma Generation and Plasma Sources. *Plasma Sources Sci. Technol.* 2000, *9* (4), 441. https://doi.org/10.1088/0963-0252/9/4/301.
20. Isbary, G.; Shimizu, T.; Li, Y.-F.; Stolz, W.; Thomas, H. M.; Morfill, G. E.; Zimmermann, J. L. Cold Atmospheric Plasma Devices for Medical Issues. *Expert Rev. Med. Devices* 2013, *10* (3), 367–377. https://doi.org/10.1586/erd.13.4.
21. Lieberman, M.; Lichtenberg, A. *Principles of Plasma Discharges and Materials Processing*, 2nd Edition; Wiley, Hoboken, NJ, 2005.
22. Rossnagel, S.; Westwood, W.; Cuomo, J. *Handbook of Plasma Processing Technology*, 1st Edition; Elsevier, Philadelphia, 1990.
23. D'Agostino, R.; Favia, P.; Fracassi, F., Eds. *Plasma Processing of Polymers*; Kluwer Academic Publishers, Dodrecht, the Netherlands, 1997.
24. Sarin, V. K. *Comprehensive Hard Materials*; Vol. 2, Elsevier, Oxford, UK, 2014.
25. Musil, J. Flexible Hard Nanocomposite Coatings. *RSC Adv.* 2015, *5* (74), 60482–60495. https://doi.org/10.1039/C5RA09586G.

26. Gu, X.; Zheng, Y.; Cheng, Y.; Zhong, S.; Xi, T. In Vitro Corrosion and Biocompatibility of Binary Magnesium Alloys. *Biomaterials* 2009, *30* (4), 484–498. https://doi.org/10.1016/j.biomaterials.2008.10.021.

27. Nichols, S. P.; Koh, A.; Storm, W. L.; Shin, J. H.; Schoenfisch, M. H. Biocompatible Materials for Continuous Glucose Monitoring Devices. *Chem. Rev.* 2013, *113* (4), 2528–2549. https://doi.org/10.1021/cr300387j.

28. Dolanský, J.; Henke, P.; Kubát, P.; Fraix, A.; Sortino, S.; Mosinger, J. Polystyrene Nanofiber Materials for Visible-Light-Driven Dual Antibacterial Action via Simultaneous Photogeneration of NO and $O_2(^1\Delta_g)$. *ACS Appl. Mater. Interfaces* 2015, *7* (41), 22980–22989. https://doi.org/10.1021/acsami.5b06233.

29. Tie, D.; Feyerabend, F.; Müller, W. D.; Schade, R.; Liefeith, K.; Kainer, K. U.; Willumeit, R. Antibacterial Biodegradable Mg-Ag Alloys. *Eur. Cell. Mater.* 2013, *25*, 284–298; discussion 298.

30. Martinho, N. Recent Advances in Drug Delivery Systems. *J. Biomater. Nanobiotechnol.* 2011, *02* (05), 510–526. https://doi.org/10.4236/jbnb.2011.225062.

31. Schoonen, L.; van Hest, J. C. M. Functionalization of Protein-Based Nanocages for Drug Delivery Applications. *Nanoscale* 2014, *6* (13), 7124–7141. https://doi.org/10.1039/C4NR00915K.

32. Naghdi, S.; Rhee, K.; Hui, D.; Park, S. A Review of Conductive Metal Nanomaterials as Conductive, Transparent, and Flexible Coatings, Thin Films, and Conductive Fillers: Different Deposition Methods and Applications. *Coatings* 2018, *8* (8), 278. https://doi.org/10.3390/coatings8080278.

33. Kim, C.-L.; Jung, C.-W.; Oh, Y.-J.; Kim, D.-E. A Highly Flexible Transparent Conductive Electrode Based on Nanomaterials. *NPG Asia Mater.* 2017, *9* (10), e438. https://doi.org/10.1038/am.2017.177.

34. Edynoor, O.; Warikh, A. R. M.; Moriga, T.; Murai, K.; Manaf, M. E. A. Transparent Coating Oxide—Indium Zinc Oxide as a Condictive Coating: A Review; *Rev. Adv. Mater. Sci.* 2017, *49*, 150–157.

35. Levchenko, I.; Keidar, M.; Mai-Prochnow, A.; Modic, M.; Cvelbar, U.; Fang, J.; Ostrikov, K. K. Plasma Treatment for Next-Generation Nanobiointerfaces. *Biointerphases* 2015, *10* (2), 029405. https://doi.org/10.1116/1.4922237.

36. Tang, L.; Wang, Y.; Li, J. The Graphene/Nucleic Acid Nanobiointerface. *Chem. Soc. Rev.* 2015, *44* (19), 6954–6980. https://doi.org/10.1039/C4CS00519H.

37. Mathur, S.; Singh, T.; Maleki, M.; Fischer, T. Plasma-Assisted Surface Treatments and Modifications for Biomedical Applications. In *Biomaterials Surface Science*; Taubert, A., F.no, J., Rodríguez-Cabello, J. C., Eds.; Wiley-VCH Verlag GmbH & Co. KGaA, 2013; pp. 375–408. https://doi.org/10.1002/9783527649600.ch13.

38. Zhao, Y.; Yeung, K. W. K.; Chu, P. K. Functionalization of Biomedical Materials Using Plasma and Related Technologies. *Appl. Surf. Sci.* 2014, *310*, 11–18. https://doi.org/10.1016/j.apsusc.2014.02.168.

39. Aliofkhazraei, M.; Ali, N. PVD Technology in Fabrication of Micro- and Nanostructured Coatings. In *Comprehensive Materials Processing*; Ed. Hashmi, S., Elsevier, Philadelphia, 2014; pp. 49–84.

40. Seshan, K. Ed. *Handbook of Thin-Film Deposition*, 3rd Edition; Noyes Publications, Norwich, NY, USA, 2012.

41. Jones, A.; Hitchman, M. Overview of Chemical Vapour Deposition. In *Chemical Vapour Deposition*; Ed. Jones, A., Hitchman, M., Royal Society of Chemistry, Cambridge, UK, 2008; pp. 1–36.

42. Martinu, L.; Zabeida, O.; Klemberg-Sapieha, J. E. Plasma-Enhanced Chemical Vapor Deposition of Functional Coatings. *Handb. Depos. Technol. Films Coat. Sci. Appl. Technol.* 2010, *445*, 392–465.

43. Gates, S. M. Surface Chemistry in the Chemical Vapor Deposition of Electronic Materials. *Chem. Rev.* 1996, *96* (4), 1519–1532. https://doi.org/10.1021/cr950233m.

44. Jasinski, J. M.; Gates, S. M. Silicon Chemical Vapor Deposition One Step at a Time: Fundamental Studies of Silicon Hydride Chemistry. *Acc. Chem. Res.* 1991, *24* (1), 9–15.

45. Adamovich, I.; Baalrud, S. D.; Bogaerts, A.; Bruggeman, P. J.; Cappelli, M.; Colombo, V.; Czarnetzki, U.; Ebert, U.; Eden, J. G.; Favia, P. et al. The 2017 Plasma Roadmap: Low Temperature Plasma Science and Technology. *J. Phys. Appl. Phys.* 2017, *50* (32), 323001. https://doi.org/10.1088/1361-6463/aa76f5.

46. Stauss, S.; Muneoka, H.; Urabe, K.; Terashima, K. Review of Electric Discharge Microplasmas Generated in Highly Fluctuating Fluids: Characteristics and Application to Nanomaterials Synthesis. *Phys. Plasmas* 2015, *22*, 057103.

47. Pamreddy, A.; Skácelová, D.; Haničinec, M.; Stahel, P.; Stupavská, M.; Černák, M.; Havel, J. Plasma Cleaning and Activation of Silicon Surface in Dielectric Coplanar Surface Barrier Discharge. *Surf. Coat. Technol.* 2013, *236*, 326–331.

48. Hess, D. W.; Reinhardt, K. A. Plasma Stripping, Cleaning, and Surface Conditioning. In *Handbook of Silicon Wafer Cleaning Technology*, Chapter 7; Ed. Reinhardt, K., Kern, W., Elsevier, New York, USA, 2018; pp. 379–455.

49. Cheruthazhekatt, S.; Černák, M.; Slavíček, P.; Havel, J. Gas Plasmas and Plasma Modified Materials in Medicine. *J. Appl. Biomed.* 2010, *8*, 55–66.

50. Ehlbeck, J.; Schnabel, U.; Polak, M.; Winter, J.; Von Woedtke, T.; Brandenburg, R.; von dem Hagen, T.; Weltmann, K. D. Low Temperature Atmospheric Pressure Plasma Sources for Microbial Decontamination. *J. Phys. D* 2011, *44*, 013002.

51. Fernández, A.; Thompson, A. The Inactivation of Salmonella by Cold Atmospheric Plasma Treatment. *Food Res. Int.* 2012, *45*, 678–684.

52. Scholtz, V.; Julak, J.; Kříha, V. The Microbicidal Effect of Low-Temperature Plasma Generated by Corona Discharge: Comparison of Various Microorganisms on an Agar Surface or in Aqueous Suspension. *Plasma Process. Polym.* 2010, *7*, 237–243.

53. Soušková, H.; Scholtz, V.; Julak, J.; Savická, D. The Fungal Spores Survival under the Low Temperature Plasma. In *NATO Science for Peace and Security Series A: Chemistry and Biology*; Ed. Machala, Z., Hensel, K., Akishev, Y., Springer, Dordrecht, 2012; pp. 57–66.

54. Alberici, S.; Dellafiore, A.; Manzo, G.; Santospirito, G.; Villa, C. M.; Zanotti, L. Organic Contamination Study for Adhesion Enhancement between Final Passivation Surface and Packaging Molding Compound. *Microelectron. Eng.* 2004, *76* (1–4), 227–234.

55. Molitor, P.; Barron, V.; Young, T. Surface Treatment of Titanium for Adhesive Bonding to Polymer Composites: A Review. *Int. J. Adhes. Adhes.* 2001, *21* (2), 129–136.

56. Bhatnagar, N. Effect of Plasma on Adhesion Characteristics of High Performance Polymers. *Rev. Adhes. Adhes.* 2013, *4*, 397–412.

57. Borooj, M. B.; Shoushtari, A. M.; Haji, A.; Sabet, E. N. Optimization of Plasma Treatment Variables for the Improvement of Carbon Fibres/Epoxy Composite Performance by Response Surface Methodology. *Compos. Sci. Technol.* 2016, *128*, 215–221.

58. Gude, M. R.; Kellar, E. J. C.; Salamat-Zadeh, F. A Reassessment of Titanium Pre-Treatments for Adhesive Bonding to Composites. *Proc. DVS-Berichte 2010 269 Conf. Join. Plast.* 2010, 99–103.

59. Ochoa-Putman, C.; Vaidya, U. K. Mechanisms of Interfacial Adhesion in Metal-Polymer Composites—Effect of Chemical Treatment. *Compos. A* 2011, *42* (8), 906–915.

60. Moosburger-Will, J.; Lachner, E.; Löffler, M.; Kunzmann, C.; Horn, S. Adhesion of Carbon Fibers to Amine Hardened Epoxy Resin: Influence of Ammonia Plasma Functionalization of Carbon Fibers. *Appl. Surf. Sci.* 2018, *453*, 141–152.

61. Zhang, Y.; Wang, Y. Y. Non-Thermal Atmospheric Plasmas in Dental Restoration: Improved Resin Adhesive Penetration. *J. Dent.* 2014, *42* (8), 1033–1042.

62. Dong, X.; Li, H.; Chen, M.; Wang, Y.; Yu, Q. Plasma Treatment of Dentin Surfaces for Improving Self-Etching Adhesive/Dentin Interface Bonding. *Clin. Plasma Med.* 2015, *3* (1), 10–16.

63. Mandolfino, C.; Lertora, E.; Genna, S.; Leone, C.; Gambaro, C. Effect of Laser and Plasma Surface Cleaning on Mechanical Properties of Adhesive Bonded Joints. *Procedia CIRP* 2015, *33*, 458–463.

64. Marchand, D. J.; Dilworth, Z. R.; Stauffer, R. J.; Hsiao, E.; Kim, J. H. Atmospheric Rf Plasma Deposition of Superhydrophobic Coatings Using Tetramethylsilane Precursor. *Surf. Coat. Technol.* 2013, *234*, 14–20.

65. Merche, D.; Vandencasteele, N.; Reniers, F. Atmospheric Plasmas for Thin Film Deposition: A Critical Review. *Thin Solid Films* 2012, *520*, 4219–4236.

66. Butscher, D.; Zimmermann, D.; Schuppler, M.; von Rohr, R. P. Plasma Inactivation of Bacterial Endospores on Wheat Grains and Polymeric Model Substrates in a Dielectric Barrier Discharge. *Food Control* 2016, *60*, 636–645.

67. Choi, S.; Puligundla, P.; Mok, C. Effect of Corona Discharge Plasma on Microbial Decontamination of Dried Squid Shreds Including Physico-Chemical and Sensory Evaluation. *LWT – Food Sci. Technol.* 2017, *75*, 323–328.

68. Devi, Y.; Thirumdas, R.; Sarangapani, C.; Deshmukh, R. R.; Annapure, U. S. Influence of Cold Plasma on Fungal Growth and Aflatoxins Production on Groundnuts. *Food Control* 2017, *77*, 187–191.

69. Sohbatzadeh, F.; Mirzanejhad, S.; Shokri, H.; Nikpour, M. Inactivation of Aspergillus Flavus Spores in a Sealed Package by Cold Plasma Streamers. *J. Theor. Appl. Phys.* 2016, *10* (2), 99–106.

70. Oh, Y. A.; Roh, S. H.; Min, S. C. Cold Plasma Treatments for Improvement of the Applicability of Defatted Soybean Meal-Based Edible Film in Food Packaging. *Food Hydrocoll.* 2016, *58*, 150–159.

71. Puligundla, P.; Lee, T.; Mok, C. Inactivation Effect of Dielectric Barrier Discharge Plasma against Foodborne Pathogens on the Surfaces of Different Packaging Materials. *Innov. Food Sci. Emerg. Technol.* 2016, *36*, 221–227.

72. Guo, J.; Huang, K.; Wang, J. Bactericidal Effect of Various Non-Thermal Plasma Agents and the Influence of Experimental Conditions in Microbial Inactivation: A Review. *Food Control* 2015, *50*, 482–490.

73. Alkawareek, M. Y.; Gorman, S. P.; Graham, W. G.; Gilmore, B. F. Potential Cellular Targets and Antibacterial Efficacy of Atmospheric Pressure Non-Thermal Plasma. *Int. J. Antimicrob. Agents* 2014, *43* (2), 154–160.

74. Misra, N. N.; Jo, C. Applications of Cold Plasma Technology for Microbiological Safety in Meat Industry. *Trends Food Sci. Technol.* 2017, *64*, 74–86.

75. Lunov, O.; Zablotskii, V.; Churpita, O.; Jager, A.; Polivka, L.; Sykova, E.; Dejneka, A.; Kubinová, S. The Interplay between Biological and Physical Scenarios of Bacterial Death Induced by Non-Thermal Plasma. *Biomaterials* 2016, *82*, 71–83.

76. Yong, H. I.; Kim, H. J.; Park, S.; Alahakoon, A. U.; Kim, K.; Choe, W.; Cheorun, J. Evaluation of Pathogen Inactivation on Sliced Cheese Induced by Encapsulated Atmospheric Pressure Dielectric Barrier Discharge Plasma. *Food Microbiol.* 2015, *46*, 46–50.

77. Liang, Y.; Wu, Y.; Sun, K.; Chen, Q.; Shen, F.; Zhang, J.; Yao, M.; Zhu, T.; Fang, J. Rapid Inactivation of Biological Species in the Air Using Atmospheric Pressure Nonthermal Plasma. *Environ. Sci. Technol.* 2012, *46* (6), 3360–3368.

78. Pankaj, S. K.; Bueno-Ferrer, C.; Misra, N. N.; Milosavljevic, V.; O'Donnell, C. P.; Bourke, P.; Keener, K. M.; Cullen, P. J. Applications of Cold Plasma Technology in Food Packaging. *Trends Food Sci. Technol.* 2014, *35*, 5–17.

79. Khaneghah, A. M.; Bagher Hashemi, S. M.; Limbo, S. Antimicrobial Agents and Packaging Systems in Antimicrobial Active Food Packaging: An Overview of Approaches and Interactions. *Food Bioprod. Process.* 2018, *111*, 1–19.

80. Asakawa, R.; Nagashima, S.; Nakamura, Y.; Hasebe, T.; Suzuki, T.; Hotta, A. Combining Polymers with Diamond-like Carbon (DLC) for Highly Functionalized Materials. *Surf. Coat. Technol.* 2011, *206*, 676–685.

81. Shirakura, A.; Nakaya, M.; Koga, Y.; Kodama, H.; Hasebe, T.; Suzuki, T. Diamond-like Carbon Films for PET Bottles and Medical Applications. *Thin Solid Films* 2006, *494*, 84–91.

82. Denes, F. S.; Manolache, S. Macromolecular Plasma-Chemistry: An Emerging Field of Polymer Science. *Prog. Polym. Sci.* 2004, *29*, 815–885.

83. Jiang, H.; Manolache, S.; Wong, A. C. L.; Denes, F. S. Plasma-Enhanced Deposition of Silver Nanoparticles onto Polymer and Metal Surfaces for the Generation of Antimicrobial Characteristics. *J. Appl. Polym. Sci.* 2004, *93*, 1411–1422.

84. George, S. M. Atomic Layer Deposition: An Overview. *Chem. Rev.* 2010, *110*, 110–131.

85. Hirvikorpi, T.; Vähä-Nissi, M.; Mustonen, T.; Iiskola, E.; Karppinen, M. Atomic Layer Deposited Aluminum Oxide Barrier Coatings for Packaging Materials. *Thin Solid Films* 2010, *518*, 2654–2658.

86. Kääriäinen, T. O.; Maydannik, P.; Cameron, D. C.; Lahtinen, K.; Johansson, P.; Kuusipalo, J. Atomic Layer Deposition on Polymer Based Flexible Packaging Materials: Growth Characteristics and Diffusion Barrier Properties. *Thin Solid Films* 2011, *519*, 3146–3154.

87. Misra, N. N.; Yepez, X.; Xu, L.; Keener, K. M. In-Package Cold Plasma Technologies. *J. Food Eng.* 2019, *244*, 21–31.

88. Heberlein, J.; Postel, O.; Girshick, S.; Mc Murry, P.; Gerberich, W.; Iordanoglou, D.; Di Fonzo, F.; Neumann, D.; Gidwani, A.; Fan, M. et al. Thermal Plasma Deposition of Nanophase Hard Coatings. *Surf. Coat. Technol.* 2001, *142–144*, 265–271.

89. Postel, O.; Heberlein, J. Deposition of Boron Carbide Thin Film by Supersonic Plasma Jet CVD with Secondary Discharge. *Surf. Coat. Technol.* 1998, *108–109*, 247–289.

90. Galevsky, G. V.; Rudneva, V. V.; Garbuzova, A. K.; Valuev, D. V. Titanium Carbide: Nanotechnology, Properties, Application. *IOP Conf. Ser. Mater. Sci. Eng.* 2015, *91*, 012017–012022.

91. Pesin, A.; Pustovoytov, D.; Vafin, R.; Yagafarov, I.; Vardanyan, E. Hardening Roll Surface by Plasma Nitriding with Subsequent Hardfacing. *J. Phys. Conf. Ser.* 2017, *830*, 012095–012101.

92. Veprek, S.; Karvankova, P.; Veprek-Heijman, M. G. J. Possible Role of Oxygen Impurities in Degradation of Nc-TiN-Si3N4 Nanocomposites. *J. Vac. Sci. Technol. B* 2005, *23*, L17–L21.

93. McIntyre, D.; Greene, J. E.; Håkansson, G.; Sundgren, J. E.; Münz, W. D. Oxidation of Metastable Single-phase Polycrystalline $Ti_{0.5}Al_{0.5}N$ Films: Kinetics and Mechanisms. *J. Appl. Phys.* 1990, *67*, 1542–1553.

94. Veprek, S.; Reiprich, S. A. Concept for the Design of Novel Superhard Coatings. *Thin Solid Films* 1995, *268* (1–2), 64–71.

95. Rebholz, C.; Schneider, J. M.; Voevodin, A. A.; Steinebrunner, J.; Charitidis, C.; Logothetidis, S.; Leyland, A.; Matthews, A. Structure, Mechanical and Tribological Properties of Sputtered TiAlBN Thin Films. *Surf. Coat. Technol.* 1999, *113* (1–2), 126–133.

96. Carvalho, S.; Rebouta, L.; Ribeiro, E.; Vaz, F.; Denannot, M. F.; Pacaud, J.; Rivière, J. P.; Paumier, F.; Gaboriaud, R. J.; Alves, E. Microstructure of (Ti,Si,Al)N Nanocomposite Coatings. *Surf. Coat. Technol.* 2004, *177-178*, 369–375.

97. Shtansky, D. V.; Sheveiko, A. N.; Petrzhik, M. I.; Kiryukhantsev-Korneev, F. V.; Levashov, E. A.; Leyland, A.; Yerokhin, A. L.; Matthews, A. Hard Tribological Ti–B–N, Ti–Cr–B–N, Ti–Si–B–N and Ti–Al–Si–B–N Coatings. *Surf. Coat. Technol.* 2005, *200* (1–4), 208–212.

98. Zhang, X.; Li, X.; Dong, H. Response of a Molybdenum Alloy to Plasma Nitriding. *Int. J. Refract. Met. Hard Mater.* 2018, *72*, 388–395.

99. Das, P.; Paul, S.; Bandyopadhyay, P. P. Plasma Sprayed Diamond Reinforced Molybdenum Coatings. *J. Alloy. Compd.* 2018, *767*, 448–455.

100. Gassner, G.; Mayrhofer, P. H.; Kutschej, K.; Mitterer, C.; Kathrein, M. Magnéli Phase Formation of PVD Mo–N and W–N Coatings. *Surf. Coat. Technol.* 2006, *201* (6), 335–3341.

101. Voevodin, A.; Muratore, C.; Zabinski, J. S. Chameleon or Smart Solid Lubricating Coatings. In *Encyclopedia of Tribology*; Ed. Wang, Q. J., Chung, Y. W., Springer, New York, USA, 2013; pp. 347–354.

102. Voevodin, A.; Muratore, C.; Aouadi, M. Hard Coatings with High Temperature Adaptive Lubrication and Contact Thermal Management: Review. *Surf. Coat. Technol.* 2014, *257*, 247–265.

103. Voevodin, A.; Jones, J. C.; Hu, J. J.; Fitz, T. A.; Zabinski, J. S. Growth and Structural Characterization of Yttria-Stabilized Zirconia-Gold Nanocomposite Films with Improved Toughness. *Thin Solid Films* 2001, *401* (1–2), 187–195.

104. Muratore, C.; Hu, J. J.; Voevodin, A. A. Tribological Coatings for Lubrication over Multiple Thermal Cycles. *Surf. Coat. Technol.* 2009, *203* (8), 957–962.

105. Schmitt, G.; Schütze, M. F.; Hays, G.; Burns, W.; Han, E. H.; Pourbaix, A.; Jacobson, G. *Global Needs for Knowledge Dissemination, Research, and Development in Materials Deterioration and Corrosion Control*. World Corrosion Organization 2009.

106. Fenker, M.; Balzer, M.; Kappl, H. Corrosion Protection with Hard Coatings on Steel: Past Approaches and Current Research Efforts. *Surf. Coat. Technol.* 2014, *257*, 182–205.

107. Fenker, M.; Balzer, M.; Kappl, H. Corrosion Behaviour of Decorative and Wear Resistant Coatings on Steel Deposited by Reactive Magnetron Sputtering—Tests and Improvements. *Thin Solid Films* 2006, *515* (1), 27–32.

108. Anders, A. A Structure Zone Diagram Including Plasma-Based Deposition and Ion Etching. *Thin Solid Films* 2010, *518* (15), 4087–4090.

109. Helmersson, U.; Lattemann, M.; Bohlmark, J.; Ehiasarian, A. P.; Gudmundsson, J. T. Ionized Physical Vapor Deposition (IPVD): A Review of Technology and Applications. *Thin Solid Films* 2006, *513* (1–2), 1–24.

110. Gudmundsson, J. T. Ionized Physical Vapor Deposition (IPVD): Magnetron Sputtering Discharges. *J. Phys. Conf. Ser.* 2008, *100* (8), 082002–082006.

111. Anders, A. Fundamentals of Pulsed Plasmas for Materials Processing. *Surf. Coat. Technol.* 2004, *183* (2–3), 301–311.

112. Lackner, J. M.; Waldhauser, W.; Ebner, R. Large-Area High-Rate Pulsed Laser Deposition of Smooth TiC_xN_{1-x} Coatings at Room Temperature—Mechanical and Tribological Properties. *Surf. Coat. Technol.* 2004, *188–189*, 519–524.

113. Gudmundsson, J. T. High Power Impulse Magnetron Sputtering Discharge. *J. Vac. Sci. Technol. A* 2012, *30*, 030801–030836.

114. Rangel, R. C.; Tagliani Pompeu, T.; Barros, J. L. S.; Antonio, C. A.; Santos, N. M.; Pelici, B. O.; Freire, C. M. A.; Cruz, N. C.; Rangel, E. C. Improvement of the Corrosion Resistance of Carbon Steel by Plasma Deposited Thin Films. In *Recent Researches in Corrosion Evaluation and Protection*; InTech, 2012; pp. 91–116.

115. Forsich, C.; Dipolt, C.; Heima, D.; Mueller, T.; Gebeshuber, A.; Holecek, R.; Lugmair, C. Potential of Thick A-C-H-Si Films as Substitute for Chromium Plating. *Surf. Coat. Technol.* 2014, *241*, 86–92.

116. Delimi, A.; Coffinier, Y.; Talhi, B.; Boukherroub, R.; Szunerits, S. Investigation of the Corrosion Protection of SiOx-like Oxide Films Deposited by PECVD onto Carbon Steel. *Electrochim. Acta* 2010, *55*, 8921–8927.

117. Wang, X. Z.; Muneshwar, T. P.; Fan, H. Q.; Cadien, K.; Luo, J. L. Achieving Ultrahigh Corrosion Resistance and Conductive Zirconium Oxynitride Coating on Metal Bipolar Plates by Plasma Enhanced Atomic Layer Deposition. *J. Pow. Sources* 2018, *397*, 32–36.

118. Barshilia, H. C.; Prakash, M. S.; Poojari, A.; Rajam, K. S. Corrosion Behavior of Nanolayered TiN/NbN Multilayer Coatings Prepared by Reactive Direct Current Magnetron Sputtering Process. *Thin Solid Films* 2004, *460* (1–2), 133–142.

119. Lang, F.; Yu, Z. The Corrosion Resistance and Wear Resistance of Thick TiN Coatings Deposited by Arc Ion Plating. *Surf. Coat. Technol.* 2001, *145* (1–3), 80–87.

120. Jehn, H. A. Improvement of the Corrosion Resistance of PVD Hard Coating–Substrate Systems. *Surf. Coat. Technol.* 2000, *125* (1–3), 212–217.

121. Aromaa, J.; Ronkainen, H.; Mahiout, A.; Hannula, S. P. Identification of Factors Affecting the Aqueous Corrosion Properties of (Ti, Al) N-Coated Steel. *Surf. Coat. Technol.* 1991, *49* (1–3), 353–358.

122. Hovsepian, P. E.; Lewis, D. B.; Luo, A.; Farinotti, A. Corrosion Resistance of CrN/NbN Superlattice Coatings Grown by Various Physical Vapour Deposition Techniques. *Thin Solid Films 488* (1–2), 1–8.

123. Lv, Y.; Ji, L.; Liu, X.; Li, H.; Zhou, H.; Chen, J. Influence of Substrate Bias Voltage on Structure and Properties of the CrAlN Films Deposited by Unbalanced Magnetron Sputtering. *Appl. Surf. Sci.* 2012, *258* (8), 3864–3870.

124. Hussain, T.; Wahab, A. A Critical Review of the Current Water Conservation Practices in Textile Wet Processing. *J Clean. Prod.* 2018, *198*, 806–819.

125. Kumar, P. S.; Narayan, A. S.; Dutta, A. *Textile Science and Clothing Technology*, Muthu, S.S. Editor.; Springer, Hong Kong, China, 2017, pp. 57–96.

126. Zille, A.; Oliveira, F. R.; Souto, A. P. Plasma Treatment in Textile Industry. *Plasma Process. Polym.* 2015, *12* (2), 98–131.

127. Chang, J. S.; Lawless, P. A.; Yamamoto, T. Corona Discharge Processes. *IEEE Trans. Plasma Sci.* 1991, *19* (6), 1152–1166.

128. Borcia, G.; Anderson, C. A.; Brown, N. M. D. Surface Treatment of Natural and Synthetic Textiles Using a Dielectric Barrier Discharge. *Surf. Coat. Technol.* 2006, *201*, 3074–3081.

129. Tyata, R. B.; Subedi, D. P.; Huczko, A. Surface Modification of Polymers and Textiles by Atmospheric Pressure Argon Glow Discharge. *Int. J. Sci. Eng. Appl. Sci.* 2016, *2* (6), 474–481.

130. Sarani, A.; Nikiforov, A.; De Geyter, N.; Morent, R.; Leys, C. Characterization of an Atmospheric Pressure Plasma Jet and Its Application for Treatment of Non-Woven Textiles. *Proc. 20th Int. Symp. Plasma Chem.* 2011.

131. Kan, C.; Lam, Y. The Effect of Plasma Treatment on Water Absorption Properties of Silk Fabrics. *Fibers Polym.* 2015, *16*, 1705–1714.

132. Vinisha Rani, K.; Chandwani, N.; Kikani, P.; Nema, S. K.; Sarma, A. K.; Sarma, B. Optimization and Surface Modification of Silk Fabric Using DBD Air Plasma for Improving Wicking Properties. *J. Text. Inst.* 2017, *109* (3), 368–375.

133. Huang, C. Y.; Wu, J. Y.; Tsai, C. S.; Hsieh, K. H.; Yeh, J. T.; Chen, K. N. Effects of Argon Plasma Treatment on the Adhesion Property of Ultra High Molecular Weight Polyethylene (UHMWPE) Textile. *Surf. Coat. Technol.* 2013, *231*, 507–511.

134. Shaw, D.; West, A.; Bredin, J.; Wagenaars, E. Mechanisms behind Surface Modification of Polypropylene Film Using an Atmospheric-Pressure Plasma Jet. *Plasma Sources Sci. Technol.* 2016, *25*, 065018–065024.

135. Zanini, S.; Barni, R.; Della Pergola, R.; Riccardi, C. Modification of the PTFE Wettability by Oxygen Plasma Treatments: Influence of the Operating Parameters and Investigation of the Ageing Behaviour. *J. Phys. D* 2014, *47* (32), 325202–325211.

136. Li, R.; Li, K.; Tian, H.; Xue, J.; Liu, S. Mechanical Properties of Plasma-treated Carbon Fiber Reinforced PTFE Composites with CNT. *Surf. Interface Anal.* 2017, *49* (11), 1064–1068.

137. Karahan, H. A.; Ozdoğan, E. Improvements of Surface Functionality of Cotton Fibers by Atmospheric Plasma Treatment. *Fibers Polym.* 2008, *9* (1), 21–26.

138. Bhat, N. V.; Netravali, A. N.; Gore, A. V.; Sathianarayanan, M. P.; Arolkar, G. A.; Deshmukh, R. R. Surface Modification of Cotton Fabrics Using Plasma Technology. *Text. Res. J.* 2011, *81* (10), 1014–1026.

139. Kan, C. W.; Lam, C. F.; Chan, C. K.; Ng, S. P. Using Atmospheric Pressure Plasma Treatment for Treating Grey Cotton Fabric. *Carbohydr. Polym.* 2014, *102* (15), 167–173.

140. Eyupoglu, S.; Kilinc, M.; Kut, D. Investigation of the Effect of Different Plasma Treatment Condition on the Properties of Wool Fabrics. *J. Text. Sci. Eng.* 2015, *5* (6), 216–222.

141. Jeon, S. H.; Hwang, K. H.; Lee, J. S.; Boo, J. H.; Yun, S. H. Plasma Treatments of Wool Fiber Surface for Microfluidic Applications. *Mater. Res. Bull.* 2015, *69*, 65–70.

142. Li, L.; Peng, M. Y.; Teng, Y.; Gao, G. Diffuse Plasma Treatment of Polyamide 66 Fabric in Atmospheric Pressure Air. *Appl. Surf. Sci.* 2016, *362* (30), 348–354.

143. Wen, Y.; Meng, X.; Liu, J.; Yan, J.; Wang, Z. Surface Modification of High-Performance Polyimide Fibers by Oxygen Plasma Treatment. *High Perform. Polym.* 2016, *29* (9), 1–7.

144. Oh, J. H.; Ko, T. J.; Moon, M. W.; Park, C. H. Nanostructured Fabric with Robust Superhydrophobicity Induced by a Thermal Hydrophobic Ageing Process. *RSC Adv.* 2017, *17*, 25597–25604.

145. Mehmood, T.; Kaynak, A. Study of Oxygen Plasma Pre-Treatment of Polyester Fabric for Improved Polypyrrole Adhesion. *Mater. Chem. Phys.*, 2014, 143 (2), 668–675.

146. Tiwari, S.; Bijwe, J. Surface Treatment of Carbon Fibers – A Review. *Proc. Technol.* 2014, *14*, 505–512.

147. Molina, J.; Fernández, J.; Fernandes, M.; Souto, A. P.; Esteves, M. F.; Bonastre, J.; Cases, F. Plasma Treatment of Polyester Fabrics to Increase the Adhesion of Reduced Graphene Oxide. *Synth. Met.* 2015, *202*, 110–122.

148. Molina, J.; Fernández, J.; Inés, J. C.; del Río, A. I.; Bonastre, J.; Cases, F. Electrochemical Characterization of Reduced Graphene Oxide-Coated Polyester Fabrics. *Electrochim. Acta* 2013, *93*, 44–52.

149. Morent, R.; De Geyter, N.; Verschuren, J.; De Clerck, K.; Kiekens, P.; Leys, C. Non-Thermal Plasma Treatment of Textiles. *Surf. Coat. Technol.* 2008, *202*, 3427–3449.

150. Paosawatyanyong, B.; Kamlangkla, K.; Hodak, S. K. Hydrophobic and Hydrophilic Surface Nano-Modification of PET Fabric by Plasma Process. *J. Nanosci. Nanotechnol.* 2010, *10* (11), 7050–7054.

151. Li, S.; Jinjin, D. Improvement of Hydrophobic Properties of Silk and Cotton by Hexafluoropropene Plasma Treatment. *Appl. Surf. Sci.* 2007, *253* (11), 5051–5055.

152. Artus, G. R. J.; Zimmermann, J.; Reifler, F. A.; Brewer, S. A.; Seeger, S. A Superoleophobic Textile Repellent towards Impacting Drops of Alkanes. *Appl. Surf. Sci.* 2012, *258* (8), 3835–3840.

153. Sun, D.; Stylios, G. K. Fabric Surface Properties Affected by Low Temperature Plasma Treatment. *J. Mater. Process. Technol.* 2006, *173* (2), 172–177.

154. Sigurdsson, S.; Shishoo, R. Surface Properties of Polymers Treated with Tetrafluoromethane Plasma. *J. Appl. Polym. Sci.* 1997, *66* (8), 1591–1601.

155. Wang, L.; Liu, D.; Xi, G. H.; Wan, S. J.; Zhao, C. H. Asymmetrically Superhydrophobic Cotton Fabrics Fabricated by Mist Polymerization of Lauryl Methacrylate. *Cellulose* 2014, *21* (4), 2983–2994.

156. Shin, B.; Lee, K. R.; Moon, M. W.; Kim, H. Y. Extreme Water Repellency of Nanostructured Low-Surface-Energy Non-Woven Fabrics. *Soft Matt.* 2012, *8* (6), 1817–1823.

157. Yasuda, H.; Yu, Q. S.; Chen, M. Interfacial Factors in Corrosion Protection: An EIS Study of Model Systems. *Prog. Org. Coat.* 2001, *41* (4), 273–279.

158. Haji, A.; Semnani Rahbar, R.; Mousavi Shoushtari, A. Plasma Assisted Attachment of Functionalized Carbon Nanotubes on Poly(Ethylene Terephthalate) Fabric to Improve the Electrical Conductivity. *Polimery* 2015, *60* (5), 337–342.

159. Wang, C. X.; Lv, J. C.; Ren, Y.; Zhi, T.; Chen, J. Y.; Zhou, Q. Q.; Lu, Z. Q.; Gao, D. W.; Jin, L. M. Surface Modification of Polyester Fabric with Plasma Pretreatment and Carbon Nanotube Coating for Antistatic Property Improvement. *Appl. Surf. Sci.* 2015, *359*, 196–203.

160. Espinosa-Cristobal, L. F.; Martinez-Castanon, G. A.; Martinez-Martinez, R. E.; Loyola-Rodriguez, J. P.; Patino-Marin, N.; Reyes-Macias, J. F.; Ruiz, F. Antibacterial Effect of Silver Nanoparticles against Streptococcus Mutans. *Mater. Lett.* 2009, *63* (29), 2603–2606.

161. Sun, G. Ed. *Antimicrobial Textiles A Volume in Woodhead Publishing Series in Textiles.* Elsevier, Duxford, UK, 2016.

162. Alongi, J.; Tata, J.; Frache, A. Hydrotalcite and Nanometric Silica as Finishing Additives to Enhance the Thermal Stability and Flame Retardancy of Cotton. *Cellulose* 2011, *18* (1), 179–190.

163. Horrocks, A. R.; Nazare, S.; Masood, R.; Kandola, B.; Price, D. Surface Modification of Fabrics for Improved Flash-fire Resistance Using Atmospheric Pressure Plasma in the Presence of a Functionalized Clay and Polysiloxane. *Polym. Adv. Technol.* 2011, *22* (1), 22–29.

164. Abou Elmaaty, T. M.; Mandour, B. A. ZnO and TiO2 Nanoparticles as Textile Protecting Agents against UV Radiation: A Review. *Asian J. Chem. Sci.* 2018, *4* (1), 1–14.

165. Jazbec, K.; Sala, M.; Mozetic, M.; Vesel, A.; Gorjanc, M. Functionalization of Cellulose Fibres with Oxygen Plasma and ZnO Nanoparticles for Achieving UV Protective Properties. *J. Nanomater.* 2015, *2015*, 346739.

166. Hench, L. L.; Polak, J. M. Third-Generation Biomedical Materials. *Science* 2002, *295*, 1014–1017.

167. Ning, C.; Zhou, L.; Tan, G. Fourth-Generation Biomedical Materials. *Mater. Today* 2016, *19* (1), 2–3.

168. Kondyurina, I.; Nechitailo, G. S.; Svistkov, A. L.; Kondyurin, A.; Bilek, M. Urinary Catheter with Polyurethane Coating Modified by Ion Implantation. *Nucl. Instr. Methods Phys. Res.* 2015, *342*, 39–46.

169. Mrad, O.; Saunier, J.; Aymes Chodur, C.; Rosilio, V.; Agnely, F.; Aubert, P.; Vigneron, J.; Etcheberry, A.; Yagoubi, N. A Comparison of Plasma and Electron Beam Sterilization of PU Catheters. *Radiat. Phys. Chem.* 2010, *79*, 93–103.

170. Mrad, O.; Saunier, J.; Aymes Chodur, C.; Mazel, V.; Rosilio, V.; Agnely, F.; Etcheberry, A.; Yagoubi, N. Aging of a Medical Device Surface Following Cold Plasma Treatment. *Eur. Polym. J.* 2011, *47* (12), 2403–2413.

171. Aflori, M.; Miron, C.; Dobromir, M.; Drobota, M. Bactericidal Effect on Foley Catheters Obtained by Plasma and Silver Nitrate Treatments. *High Perform. Polym.* 2015, *27*, 655–660.

172. Zare, H. H.; Juhart, V.; Vass, A.; Franz, G.; Jocham, D. Efficacy of Silver/Hydrophilic Poly(p-Xylylene) on Preventing Bacterial Growth and Biofilm Formation in Urinary Catheters. *Biointerphases* 2017, *12* (1), 011001–011010.

173. Mendoza, G.; Regiel-Futyra, A.; Tamayo, A.; Monzon, M.; Irusta, S.; de Gregorio, M. A.; Kyziol, A.; Arruebo, M. Chitosan-Based Coatings in the Prevention of Intravascular Catheter-Associated Infections. *J. Biomater. Appl.* 2018, *32* (6), 725–737.

174. Tyler, B. J.; Hook, A.; Pelster, A.; Williams, P.; Morgan, A.; Arlinghaus, H. F. Development and Characterization of a Stable Adhesive Bond between a Poly(Dimethylsiloxane) Catheter Material and a Bacterial Biofilm Resistant Acrylate Polymer Coating. *Biointerphases* 2017, *12* (2), 02C412–02C424.

175. Badv, M.; Jaffer, I. H.; Weitz, J. I.; Didar, T. F. An Omniphobic Lubricant-Infused Coating Produced by Chemical Vapor Deposition of Hydrophobic Organosilanes Attenuates Clotting on Catheter Surfaces. *Sci. Rep.* 2017, *7*, 11639–11649.

176. Zheng, Y.; Miao, J.; Zhang, F.; Cai, C.; Koh, A.; Simmons, T. J.; Mousa, S. A.; Linhardtb, R. J. Surface Modification of a Polyethylene Film for Anticoagulant and Anti-Microbial Catheter. *React. Funct. Polym.* 2016, *100*, 142–150.

177. Thakur, S.; Pal, D.; Neogi, S. Prevention of Biofilm Attachment by Plasma Treatment of Polyethylene. *Surf. Innovation*, 41, 33–38.

178. Kochkodan, V. M.; Sharma, V. K. Graft Polymerization and Plasma Treatment of Polymer Membranes for Fouling Reduction: A Review. *J. Environ. Sci. Health Tox Hazard Subst. Environ. Eng.* 2012, *47* (12), 1713–1727.

179. Kim, K. S.; Lee, K. H.; Cho, K.; Park, C. E. Surface Modification of Polysulfone Ultrafiltration Membrane by Oxygen Plasma Treatment. *J. Membr. Sci.* 2002, *199*, 135–145.

180. He, X. C.; Yu, H. Y.; Tang, Z. Q.; Liu, L. Q.; Yan, M. G.; Gu, J. S.; Wei, X. W. Reducing Protein Fouling of a Polypropylene Microporous Membrane by CO_2 Plasma Surface Modification. *Desalination* 2009, *244*, 80–89.

181. Kull, K. R.; Steen, M. L.; Fisher, E. R. Surface Modification with Nitrogen-Containing Plasmas to Produce Hydrophilic, Low-Fouling Membranes. *J. Membr. Sci.* 2005, *246*, 203–215.

182. Watkins, L. M.; Lee, A. F.; Moir, J. W. B.; Wilson, K. Plasma-Generated Poly(Allyl Alcohol) Antifouling Coatings for Cellular Attachment. *ACS Biomater. Sci. Eng.* 2017, *3* (1), 88–94.

183. Pandiyaraj, K. N.; Ramkumar, M. C.; Arun Kumar, A.; Padmanabhan, P. V. A.; Pichumani, M.; Bendavid, A.; Cools, P.; De Geyter, N.; Morent, R.; Kumar, V. et al. Evaluation of Surface Properties of Low Density Polyethylene (LDPE) Films Tailored by Atmospheric Pressure Non-Thermal Plasma (APNTP) Assisted Co-Polymerization and Immobilization of Chitosan for Improvement of Antifouling Properties. *Mater. Sci. Eng. C* 2019, *94*, 150–160.

184. Pandiyaraj, K. N.; Arun Kumar, A.; Ramkumar, M. C.; Padmanabhan, P. V. A.; Trimukhe, A. M.; Deshmukh, R. R.; Cools, P.; Morent, R.; De Geyter, N.; Kumar, V. et al. Influence of Operating Parameters on Development of Polyethylene Oxide-like Coatings on the Surfaces of Polypropylene Films by Atmospheric Pressure Cold Plasma Jet-Assisted Polymerization to Enhance Their Antifouling Properties. *J. Phys. Chem. Sol.* 2018, *123*, 76–86.

185. Pandiyaraj, K. N.; Deshmukh, R. R.; Arunkumar, A.; Ramkumar, M. C.; Ruzybayev, I.; Shah, S. I.; Su, P. G.; Halleluyah, M. J.; Halim, A. S. B. Evaluation of Mechanism of Non-Thermal Plasma Effect on the Surface of Polypropylene Films for Enhancement of Adhesive and Hemo Compatible Properties. *Appl. Surf. Sci.* 2015, *347*, 336–346.

186. Hsiao, C. R.; Lin, C. W.; Chou, C. M.; Chung, C. J.; He, J. L. Surface Modification of Blood-Contacting Biomaterials by Plasma-Polymerized Superhydrophobic Films Using Hexamethyldisiloxane and Tetrafluoromethane as Precursors. *Appl. Surf. Sci.* 2015, *346*, 50–56.

187. Alves, P.; Cardoso, R.; Correia, T. R.; Antunes, B. P.; Correia, I. J.; Ferreira, P. Surface Modification of Polyurethane Films by Plasma and Ultraviolet Light to Improve Hemocompatibility for Artificial Heart Valves. *Coll. Surf. B* 2014, *113*, 25–32.

188. Solouk, A.; Cousins, B. G.; Mirahmadi, F.; Mirzadeh, H.; Nadoushan, M. R.; Shokrgozar, M. A.; Seifalian, A. M. Biomimetic Modified Clinical-Grade POSS-PCU Nanocomposite Polymer for Bypass Graft Applications: A Preliminary Assessment of Endothelial Cell Adhesion and Haemocompatibility. *Mater. Sci. Eng. C* 2015, *46*, 400–408.

189. Lukas, K.; Thomas, U.; Gessner, A.; Wehner, D.; Schmid, T.; Schmid, C.; Lehle, K. Plasma Functionalization of Polycarbonaturethane to Improve Endothelialization-Effect of Shear Stress as a Critical Factor for Biocompatibility Control. *J. Biomater. Appl.* 2016, *30* (9), 1417–1428.

190. Chen, S. H.; Chang, Y.; Lee, K. R.; Wei, T. C.; Higuchi, A.; Ho, F. M.; Tsou, C. C.; Ho H.T.; Lai, Y. T. Hemocompatible Control of Sulfobetaine-Grafted Polypropylene Fibrous Membranes in Human Whole Blood via Plasma-Induced Surface Zwitterionization. *Langmuir* 2012, *28* (51), 17733–17742.

191. Chang, Y.; Chang, Y.; Higuchi, A.; Shih, Y. J.; Li, P. T.; Chen, W. Y.; Tsai, E. M.; Hsiue, G. H. Bioadhesive Control of Plasma Proteins and Blood Cells from Umbilical Cord Blood onto the Interface Grafted with Zwitterionic Polymer Brushes. *Langmuir* 2012, *28* (9), 4309–4317.

192. Chang, Y.; Shih, Y. J.; Ko, C. Y.; Jhong, J. F.; Liu, Y. L.; Wei, T. C. Hemocompatibility of Poly(Vinylidene Fluoride) Membrane Grafted with Network-Like and Brush-Like Antifouling Layer Controlled via Plasma-Induced Surface PEGylation. *Langmuir* 2011, *27* (9), 5445–5455.

193. de Mel, A.; Cousins, B. G.; Seifalian, A. M. Surface Modification of Biomaterials: A Quest for Blood Compatibility. *Int. J. Biomater.* 2012, *2012*, 707863–707871.

194. McGuigan, A. P.; Sefton, M. V. The Influence of Biomaterials on Endothelial Cell Thrombogenicity. *Biomaterials* 2007, *28* (16), 2547–2571.

195. Jeon, H. J.; Lee, H.; Kim, G. H. Nano-Sized Surface Patterns on Electrospun Microfibers Fabricated Using a Modified Plasma Process for Enhancing Initial Cellular Activities. *Plasma Proc. Polym.* 2013, *11*, 142–148.

196. Sharifi, F.; Irani, S.; Zandi, M.; Soleimani, M.; Atyabi, S. M. Comparative of Fibroblast and Osteoblast Cells Adhesion on Surface Modified Nanofibrous Substrates Based on Polycaprolactone. *Prog. Biomater.* 2016, *5*, 213–222.

197. Goldman, M.; Juodzbalys, G.; Vilkinis, V. Titanium Surfaces with Nanostructures Influence on Osteoblasts Proliferation: A Systematic Review. *J. Oral Maxillofac. Res.* 2014, *5* (3), e1–e14.

198. Martins, A.; Chung, S.; Pedro, A. J.; Sousa, R. A.; Marques, A. P.; Reis, R. L.; Neves, N. M. Hierarchical Starch-Based Fibrous Scaffold for Bone Tissue Engineering Applications. *J. Tissue Eng. Regen. Med.* 2009, *3*, 37–42.

199. Kubo, K.; Tsukimura, N.; Iwasa, F.; Ueno, T.; Saruwatari, L.; Aita, H.; Chiou, W. A.; Ogawa, T. Cellular Behavior on TiO2 Nanonodular Structures in a Micro-to-Nanoscale Hierarchy Model. *Biomaterials* 2009, *30*, 5319–5329.

200. Cheng, Q.; Lee, B. L. P.; Komvopoulos, K.; Yan, Z.; Li, S. Plasma Surface Chemical Treatment of Electrospun Poly(L-Lactide) Microfibrous Scaffolds for Enhanced Cell Adhesion, Growth, and Infiltration. *Tissue Eng. A* 2013, *19* (9–10), 1188–1198.

201. Floren, M.; Bonani, W.; Dharmarajan, A.; Motta, A.; Migliaresi, C.; Tan, W. Human Mesenchymal Stem Cells Cultured on Silk Hydrogels with Variable Stiffness and Growth Factor Differentiate into Mature Smooth Muscle Cell Phenotype. *Acta Biomater.* 2016, *31*, 156–166.

202. Foss, C.; Merzari, E.; Migliaresi, C.; Motta, A. Silk Fibroin/ Hyaluronic Acid 3D Matrices for Cartilage Tissue Engineering. *Biomacromolecules* 2013, *14* (1), 38–47.

203. Silva, S. S.; Maniglio, D.; Motta, A.; Mano J. F.; Migliaresi, C. Genipin-Modified Silk-Fibroin Nanometric Nets. *Macromol. Biosci.* 2008, *8* (8), 766–774.

204. Hodgkinson, T.; Yuan, X. F.; Bayat, A. Electrospun Silk Fibroin Fiber Diameter Influences in Vitro Dermal Fibroblast Behavior and Promotes Healing of Ex Vivo Wound Models. *J. Tissue Eng.* 2014, *5*, 1–13.

205. Jeong, L.; Yeo, I. S.; Kim, H. M.; Yoon, Y. I.; Jang, D. H.; Jung, S. Y.; Min, B. M.; Park, W. H. Plasma-Treated Silk Fibroin Nanofibers for Skin Regeneration. *Int. J. Biol. Macromol.* 2009, *44* (3), 222–228.

206. Savoji, H.; Maire, M.; Lequoy, P.; Liberelle, B.; De Crescenzo, G.; Ajji, A.; Wertheimer, M. R.; Lerouge, S. *Biomacromolecules* 2017, *18* (1), 303–310.

207. Abdelrahim, R. A.; Badr, N. A.; Baroudi, K. The Effect of Plasma Surface Treatment on the Bioactivity of Titanium Implant Materials (in Vitro). *J. Int. Soc. Prev. Community Dent.* 2016, *6* (1), 15–21.

208. Gkomoza, P.; Lampropoulos, G. S.; Vardavoulias, M.; Pantelis, D. I.; Karakizis, P. N.; Sarafoglou, C. Microstructural Investigation of Porous Titanium Coatings, Produced by Thermal Spraying Techniques, Using Plasma Atomization and Hydride-Dehydride Powders, for Orthopedic Implants. *Surf. Coat. Technol.* 2019, *357*, 947–956.

209. de Freitas Daudt, N.; Bram, M.; Cysne Barbosa, A. P.; Laptev, A. M.; Alves Jr., C. Manufacturing of Highly Porous Titanium by Metal Injection Molding in Combination with Plasma Treatment. *J. Mater. Process. Technol.* 2017, *239*, 202–209.

210. Chou, W. C.; Wang, R. C. C.; Huang, C. L.; Lee, T. M. The Effect of Plasma Treatment on the Osseointegration of Rough Titanium Implant: A Histo-Morphometric Study in Rabbits. *J. Dent. Sci.* 2018, *13* (3), 267–273.

211. Choi, Y. R.; Kwon, J. S.; Song, D. H.; Choi, E. H.; Lee, Y. K.; Kim, K. N.; Kim, K. M. Surface Modification of Biphasic Calcium Phosphate Scaffolds by Non-Thermal Atmospheric Pressure Nitrogen and Air Plasma Treatment for Improving Osteoblast Attachment and Proliferation. *Thin Solid Films* 2013, *547*, 235–240.

212. Bergemann, C.; Cornelsen, M.; Quade, A.; Laube, T.; Schnabelrauch, M.; Rebl, H.; Weißmann, V.; Seitz, H.; Nebe, B. Continuous Cellularization of Calcium Phosphate Hybrid Scaffolds Induced by Plasma Polymer Activation. *Mater. Sci. Eng. C* 2016, *59*, 514–523.

213. Roh, H. S.; Lee, C. M.; Hwang, Y. H.; Kook, M. S.; Yang, S. W.; Lee, D.; Kim, B. H. Addition of MgO Nanoparticles and Plasma Surface Treatment of Three-Dimensional Printed Polycaprolactone/ Hydroxyapatite Scaffolds for Improving Bone Regeneration. *Mater. Sci. Eng. C* 2017, *74*, 525–535.

214. Bos, G. W.; Scharenborg, N. M.; Poot, A. A.; Engbers, G. H.; Beugeling, T.; van Aken, W. G.; Feijen, J. Blood Compatibility of Surfaces with Immobilized Albumin-Heparin Conjugate and Effect of Endothelial Cell Seeding on Platelet Adhesion. *J. Biomed. Mater. Res.* 1999, *47* (3), 279–291.

215. Dekker, A.; Reitsma, K.; Beugeling, T.; Bantjes, A.; Feijen, J.; van Aken, W. G. Adhesion of Endothelial Cells and Adsorption of Serum Proteins on Gas Plasma-Treated Polytetrafluoroethylene. *Biomaterials* 1991, *12* (2), 130–138.

216. Nik, N. S.; Amoabediny, G.; Shokrgozar, M. A.; Mottaghy, K.; Klein-Nulend, J.; Zandieh-Doulabi, B. Surface Modification of Silicone Tubes by Functional Carboxyl and Amine, but Not Peroxide Groups Followed by Collagen Immobilization Improves Endothelial Cell Stability and Functionality. *Biomed. Mater.* 2010, *10*, 015024–015034.

217. Reznickova, A.; Novotna, Z.; Kolska, Z.; Kasalkova, N. S.; Rimpelova, S.; Svorcik, V. Enhanced Adherence of Mouse Fibroblast and Vascular Cells to Plasma Modified Polyethylene. *Mater. Sci. Eng. C* 2015, *52*, 259–266.

218. Nedela, O.; Slepicka, P.; Kolska, Z.; Slepickov, K.; Kasalkova, N. S.; Sajdl, P.; Vesely, M.; Svorcik, V. Functionalized Polyethylene Naphthalate for Cytocompatibility Improvement. *React. Funct. Polym.* 2016, *100*, 44–52.

219. Biazar, E.; Heidari, M.; Asefnezhad, A.; Montazeri, N. The Relationship between Cellular Adhesion and Surface Roughness in Polystyrene Modified by Microwave Plasma Radiation. *Int. J. Nanomed.* 2011, *6*, 631–639.

220. Richards, R. G. The Effect of Surface Roughness on Fibroblast Adhesion in Vitro. *Injury* 1996, *27* (3), S/C38–S/C43.

221. Huang, H. H.; Ho, C. T.; Lee, T. H.; Lee, T. L.; Liao, K. K.; Chen, F. L. Effect of Surface Roughness of Ground Titanium on Initial Cell Adhesion. *Biomol. Eng.* 2004, *21* (3–5), 93–97.

222. Hallab, N. J.; Bundy, K. J.; O'Connor, K.; Clark, R.; Moses, R. L. Cell Adhesion to Biomaterials: Correlations between Surface Charge, Surface Roughness, Adsorbed Protein, and Cell Morphology. *J. Long Term Eff. Med. Implants* 1995, *5* (3), 209–231.

223. Antonini, V.; Torrengo, S.; Marocchi, L.; Minati, L.; Dalla Serra, M.; Bao, G.; Speranza, G. Combinatorial Plasma Polymerisation Approach to Produce Thin Films for Testing Cell Proliferation. *Coll. Surf. B* 2013, *113*, 320–329.

224. Hollister, S. J. Porous Scaffold Design for Tissue Engineering. *Nat. Mater.* 2005, *4* (7), 518–524.

225. Dhandayuthapani, B.; Yoshida, Y.; Maekawa, T.; Kumar, D. S. Polymeric Scaffolds in Tissue Engineering Application: A Review. *Int. J. Polym. Sci.* 2011, *2011*, 1–19.

226. Tanner, K. Bioactive Composites for Bone Tissue Engineering. Proceedings of the Institution of Mechanical Engineers. Part H. *J. Eng. Med.* 2010, *224* (12), 1359–1372.

227. O'Brien, J. Biomaterials & Scaffolds for Tissue Engineering. *Mater. Today* 2011, *14* (3), 88–95.

228. Gleadall, A.; Visscher, D.; Yang, J.; Thomas, D.; Segal, J. Review of Additive Manufactured Tissue Engineering Scaffolds: Relationship between Geometry and Performance. *Burns Trauma* 2018, *6*, 19–35.

229. Ghassemi, T.; Shahroodi, A.; Ebrahimzadeh, M. H.; Mousavian, A.; Movaffagh, J.; Moradi, A. Current Concepts in Scaffolding for Bone Tissue Engineering. *Arch. Bone Joint Surg.* 2018, *6* (2), 90–99.

230. Jacobs, T.; Morent, R.; De Geyter, N.; Dubruel, P.; Leys, C. Plasma Surface Modification of Biomedical Polymers: Influence on Cell-Material Interaction. *Plasma Chem. Plasma Process.* 2012, *32* (5), 1039–1073.

231. Yang, J.; Shi, G.; Wang, S.; Cao, Y.; Shang, Q.; Yang, G.; Wang, W. Fabrication and Surface Modification of Macroporous Poly(L-Lactic Acid) and Poly(L-Lactic-Co-Glycolicacid) (70/30) Cell Scaffolds for Human Skin Fibroblast Cell Culture. *J. Biomed. Mater. Res.* 2002, *62* (3), 438–446.

232. Choi, Y. R.; Kwon, J. S.; Song, D. H.; Choi, E. H.; Lee, Y. K.; Kim, K. N.; Kim, K. M. Surface Modification of Biphasic Calcium Phosphate Scaffolds by Non-Thermal Atmospheric Pressure Nitrogen and Air Plasma Treatment for Improving Osteoblast Attachment and Proliferation. *Thin Solid Films* 2013, *547*, 235–240.

233. Ring, A.; Langer, S.; Schaffran, A.; Stricker, I.; Awakowicz, P.; Steinau, H. U.; Hauser, J. Enhanced Neovascularization of Dermis Substitutes via Low-Pressure Plasma-Mediated Surface Activation. *Burns* 2010, *36* (8), 1222–1227.

234. Shah, A. R.; Cornejo, A.; Guda, T.; Sahar, D. E.; Stephenson, S. M.; Chang, S.; Krishnegowda, N. K.; Sharma, R.; Wang, H. T. Differentiated Adipose-Derived Stem Cell Cocultures for Bone Regeneration in Polymer Scaffolds in Vivo. *J. Craniofac. Surg.* 2014, *25* (4), 1504–1509.

235. Sorrell, J. M. A Self-Assembled Fibroblast-Endothelial Cell Co-Culture System That Supports in Vitro Vasculogenesis by Both Human Umbilical Vein Endothelial Cells and Human Dermal Microvascular Endothelial Cells. *Cells Tissues Organs* 2007, *186*, 157–168.

236. Chen, R. R.; Silva, E. A.; Yuen, W. W.; Mooney, D. J. Spatio-Temporal VEGF and PDGF Delivery Patterns Blood Vessel Formation and Maturation. *Pharm. Res.* 2007, *24*, 258–264.

237. Cools, P.; Mota, C.; Moldero, I. L.; Ghobeira, R.; De Geyter, N.; Moroni, L.; Morent, R. Acrylic Acid Plasma Coated 3D Scaffolds for Cartilage Tissue Engineering Applications. *Sci. Rep.* 2018, *8*, 3830–3845.

238. Yan, H.; Yuanhao, W.; Hongxing, Y. TEOS/Silane Coupling Agent Composed Double Layers Structure: A Novel Super-Hydrophilic Coating with Controllable Water Contact Angle Value. *Appl. Energy* 2017, *185*, 2209–2216.

239. Parkin, I. P.; Palgrave, R. G. Self-Cleaning Coatings. *J. Mater. Chem.* 2005, *15*, 1689–1695.

240. Hunter, S. R.; Smith, D. B.; Polizos, G.; Schaeffer, D. A.; Lee, D. F.; Datskos, P. G. Low Cost Anti-Soiling Coatings for CSP Collector Mirrors and Heliostats, in: High and Low Concentrator Systems for Solar Energy Applications. *Proc. High Low Conc. Syst. Sol. Energy Appl. IX Int. Soc. Opt. Photonics* 2014, 91750J.

241. Midtdal, K.; Jelle, B. P. Self-Cleaning Glazing Products: A State-of-the-Art Review and Future Research Pathways. *Sol. Energy Mater. Sol. Cells* 2013, *109*, 126–141.

242. von Niessen, K.; Gindrat, M. Plasma Spray-PVD, A New Thermal Spray Process to Deposit out of the Vapor Phase. *J. Therm. Spray Technol.* 2011, *20*, 736–743.

243. Mozumder, M. S.; Mourad, A. H. I.; Pervez, H.; Surkatti, R. Recent Developments in Multifunctional Coatings for Solar Panel Applications: A Review. *Sol. Energy Mater. Sol. Cells* 2019, *189*, 75–102.

244. Adak, D.; Ghosh, S.; Chakraborty, P.; Srivatsa, K. M. K.; Mondal, A. Non Lithographic Block Copolymer Directed Self-Assembled and Plasma Treated Self-Cleaning Transparent Coating for Photovoltaic Modules and Other Solar Energy Devices. *Sol. Energy Mater. Sol. Cells* 2018, *188*, 127–139.

245. Prabhu, S.; Cindrella, L.; Joong Kwon, O.; Mohanraju, K. Superhydrophilic and Selfcleaning RGO-TiO_2 Composite Coatings for Indoor and Outdoor Photovoltaic Applications. *Sol. Energy Mater. Sol. Cells* 2017, *169*, 304–312.

246. Jesus, M. A. M. L.; Neto, J. T. S.; Timo, G.; Paiva, P. R. P.; Dantas, M. S. S.; Ferreira, A. M. Superhydrophilic Self-Cleaning Surfaces Based on TiO2 and TiO2/SiO2 Composite Films for Photovoltaic Module Cover Glass. *Appl. Adhes. Sci.* 2015, *3*, 5–14.

247. Zardetto, V.; Di Giacomo, F.; Lucarelli, G.; Kessels, W. M. M.; Brown, T. M.; Creatore, M. Plasma-Assisted Atomic Layer Deposition of TiO_2 Compact Layers for Flexible Mesostructured Perovskite Solar Cells. *Sol. Energy* 2017, *150* (1), 447–453.

248. Sung, Y. M. Deposition of TiO2 Blocking Layers of Photovoltaic Cell Using RF Magnetron Sputtering Technology. *Energy Procedia* 2013, *34*, 582–588.

249. Chou, W. C.; Liu, W. J. Study of Dye Sensitized Solar Cell Application of TiO_2 Films by Atmospheric Pressure Plasma Deposition Method. *Proc. Int. Conf. Electron. Packag. ICEP* 2016, 664–668.

250. Filippin, A. N.; Sanchez-Valencia, J. R.; Idígoras, J.; Rojas, T. C.; Barranco, A.; Anta, J. A.; Borras, A. Plasma Assisted Deposition of Single and Multistacked TiO_2 Hierarchical Nanotube Photoanodes. *Nanoscale* 2017, *9*, 8133–8141.

251. Swatowska, B.; Stapinski, T.; Drabczkyk, K.; Panek, P. The Role of Antireflective Coatings in Silicon Solar Cells the Influence on Their Electrical Parameters. *Opt. App.* 2011, *41* (2), 487–492.

252. Sharma, N.; Kumar, M.; Kumari, N.; Karar, V.; Sharma, A. L. Design and Deposition of Single and Multilayer Antireflection Coatings of Glass Substrate Using Electron Beam Deposition. *Mater. Today Proc.* 2018, *5* (2), 6421–6425.

253. Selj, J. K.; Young, D.; Grover, S. Optimization of the Antireflection Coating of Thin Epitaxial Crystalline Silicon Solar Cells. *Energy Procedia* 2015, *77*, 248–252.

254. Mehmood, U.; Al-Sulaiman, F. A.; Yilbas, B. S.; Salhi, B.; Ahmed, S. H. A.; Hossain, M. K. Superhydrophobic Surfaces with Antireflection Properties for Solar Applications: A Critical Review. *Sol. Energy Mater. Sol. Cells* 2016, *157*, 604–623.

255. Ganesh, V. A.; Nairb, A. S.; Ramakrishna, S. Anti-Reflective Coatings: A Critical, in-Depth Review. *Energy Environ. Sci.* 2011, *4*, 3779–3803.

256. Moghal, J.; Kobler, J.; Sauer, J.; Best, J.; Gardener, M.; Watt, A.; Wakefield, G. High-Performance, Single-Layer Antireflective Optical Coatings Comprising Mesoporous Silica Nanoparticles. *ACS Appl. Mater. Interfaces* 2012, *4*, 854–859.

257. Xu, L.; Gao, L.; He, L. Fabrication of Visible/near-IR Antireflective and Superhydrophobic Coatings from Hydrophobically Modified Hollow Silica Nanoparticles and Poly (Methyl Methacrylate). *RSC Adv.* 2012, *2*, 12764–12769.

258. Xiao, S. Q.; Xu, S.; Gu, X. F.; Song, D. Y.; Zhoub, H. P.; Ostrikov, K. Chemically Active Plasmas for Surface Passivation of Si Photovoltaics. *Catal. Today* 2015, *252*, 201–210.

259. Ali, K.; Khan, S. A.; Mat Jafri, M. Z. Effect of Double Layer (SiO$_2$/TiO$_2$) Anti-Reflective Coating on Silicon Solar Cells. *Int. J. Electrochem. Sci.* 2014, *9*, 7865–7874.

260. Richards, B. S.; Rowlands, S. F.; Honsberg, C. B.; Cotter, J. E. TiO$_2$ DLAR Coatings for Planar Silicon Solar Cells. *Prog. Photovoltaics* 2003, *11* (1), 27–32.

261. Womack, G.; Kaminski, P. M.; Abbas, A.; Isbilir, K.; Gottschalg, R.; Walls, J. M. Performance and Durability of Broadband Antireflection Coatings for Thin Film CdTe Solar Cells. *J. Vac. Sci. Technol. A* 2017, *35*, 021201–021212.

262. Tai, Q.; Yan, F. Emerging Semitransparent Solar Cells: Materials and Device Design. *Adv. Mater.* 2017, *29* (34), 1700192–100226.

263. Mrázková, Z.; Postava, K.; Torres-Rios, A.; Foldyna, M.; Cabarrocas, P. R.; Pištora, J. Optical Modeling of Microcrystalline Silicon Deposited by Plasma-Enhanced Chemical Vapor Deposition on Low-Cost Iron-Nickel Substrates for Photovoltaic Applications. *Procedia Mater. Sci.* 2016, *12*, 130–135.

264. Long, J.; Yin, Y.; Sian, S. Y. R.; Ren, Z.; Wang, J.; Vayalakkara, P.; Venkataraj, S.; Aberle, A. G. Doped Microcrystalline Silicon Layers for Solar Cells by 13.56MHz Plasma-Enhanced Chemical Vapour Deposition. *Energy Procedia* 2012, *15*, 240–247.

265. Roschek, T.; Repmann, T.; Müller, J.; Rech, B.; Wagner, H. Comprehensive Study of Microcrystalline Silicon Solar Cells Deposited at High Rate Using 13.56 MHz Plasma-Enhanced Chemical Vapor Deposition. *J. Vac. Sci. Technol. A* 2001, *20*, 492–498.

266. Schmidt, J.; Peibst, R.; Brendel, R. Surface Passivation of Crystalline Silicon Solar Cells: Present and Future. *Sol. Energy Mater. Sol. Cells* 2018, *187*, 39–54.

267. Pikna, P.; Skoromets, V.; Becker, C.; Fejfar, A.; Kužel, P. Thin Film Polycrystalline Silicon Solar Cells Studied by Transient Terahertz Probe Spectroscopy. *Energy Procedia* 2016, *102*, 19–26.

268. Schmidt, J.; Werner, F.; Veith, B.; Zielke, D.; Steingrube, S.; Altermatt, P. P.; Gatz, S.; Dullweber, T.; Brendel, R. Advances in the Surface Passivation of Silicon Solar Cells. *Energy Procedia* 2012, *15*, 30–39.

269. Patel, M.; Kim, H. S.; Kim, J.; Yun, J. H.; Kim, S. J.; Choi, E. H.; Park, H. H. Excitonic Metal Oxide Heterojunction (NiO/ZnO) Solar Cells for All-Transparent Module Integration. *Sol. Energy Mater. Sol. Cells* 2017, *170*, 246–253.

270. Stoklas, R.; Gregušová, D.; Hasenöhrl, S.; Brytavskyi, E.; Ťapajna, M.; Fröhlich, K.; Haščík, S.; Gregor, M.; Kuzmík, J. Characterization of Interface States in AlGaN/GaN Metal-Oxide-Semiconductor Heterostructure Field-Effect Transistors with HfO$_2$ Gate Dielectric Grown by Atomic Layer Deposition. *Appl. Surf. Sci.* 2018, *461*, 255–259.

271. Kato, Y.; Jung, M. C.; Lee, M. V.; Qi, Y. Electrical and Optical Properties of Transparent Flexible Electrodes: Effects of UV Ozone and Oxygen Plasma Treatments. *Org. Electron.* 2014, *15* (3), 721–728.

272. Luo, Y.; Cheng, R.; Shen, J.; Chen, X.; Lu, Z.; Chen, Y.; Sun, Z.; Huang, S. Plasma-Modified SnO2:F Substrate for Efficient Cobalt Selenide Counter in Dye Sensitized Solar Cell. *RSC Adv.* 2014, *4*, 44896–44901.

273. Poortmans, J. Epitaxial Thin Film Crystalline Silicon Solar Cells on Low Cost Silicon Carriers. In *Thin Film Solar Cells Fabrication, Characterization and Applications*; Ed. Poortmans, J., Arkhipov, V., Wiley, Chichester, UK, 2006.

274. De Wolf, S.; Descoeudres, A.; Holman, Z. C.; Ballif, C. High-Efficiency Silicon Heterojunction Solar Cells: A Review. *Green* 2012, *2*, 7–24.

275. Yu, C.; Yang, M.; Dong, G.; Peng, F.; Hu, D. C.; Long, W.; Hong, C.; Cui, G.; Wang, J.; He, Y. et al. Development of Silicon Heterojunction Solar Cell Technology for Manufacturing. *Jpn. J. Appl. Phys.* 2018, *57*, 08RB15–08RB23.

276. Nakano, Y. Ultra-High Efficiency Photovoltaic Cells for Large Scale Solar Power Generation. *AMBIO* 2012, *41* (2 Supplement), 125–131.

277. Xu, X.; Zhang, J.; Hu, A.; Yu, C.; Qu, M.; Peng, C.; Ru, X.; Wang, J.; Lin, F.; Shan, H. et al. Development of Nanocrystalline Silicon Based Multi-Junction Solar Cell Technology for High Volume Manufacturing. *Mater. Res. Soc. Symp. Proc.* 2013, 1536.

278. Hamon, G.; Vaissiere, N.; Cariou, R.; Lachaume, R.; Alvarez, J.; Chen, W.; Kleider, J. P.; Decobert, J.; Cabarrocas, P. R. Plasma-Enhanced Chemical Vapor Deposition Epitaxy of Si on GaAs for Tunnel Junction Applications in Tandem Solar Cells. *J. Photon. Energy* 2017, *7* (2), 022504–022512.

279. Tommi, T.; Maarit, K. Atomic Layer Deposition of ZnO: A Review. *Semicond. Sci. Technol.* 2014, *29* (4), 043001–043016.

280. Hausmann, D. M.; Kim, E.; Becker, J.; Gordon, R. G. Atomic Layer Deposition of Hafnium and Zirconium Oxides Using Metal Amide Precursors. *Chem. Mater.* 2002, *14*, 4350–4358.

281. Hämäläinen, J.; Ritala, M.; Leskelä, M. Atomic Layer Deposition of Noble Metals and Their Oxides. *Chem. Mater.* 2014, *26*, 786–801.

282. Song, G. Y.; Oh, C.; Sinha, S.; Son, J.; Heo, J. Facile Phase Control of Multivalent Vanadium Oxide Thin Films (V_2O_5 and VO_2) by Atomic Layer Deposition and Post-Deposition Annealing. *ACS Appl. Mater. Interfaces* 2017, *9*, 23909–23917.

283. Dasgupta, N. P.; Meng, X.; Elam, J. W.; Martinson, A. B. F. Atomic Layer Deposition of Metal Sulfide Materials. *Acc. Chem. Res.* 2015, *48*, 341–348.

284. Sinha, S.; Mahuli, N.; Sarkar, S. K. Atomic Layer Deposition of Aluminum Sulfide Thin Films Using Trimethylaluminum and Hydrogen Sulfide. *J. Vac. Sci. Technol. A* 2015, *33*, 01A139–01A147.

285. Nandi, D. K.; Sen, U. K.; Dhara, A.; Mitra, S.; Sarkar, S. K. Intercalation Based Tungsten Disulfide (WS_2) Li-Ion Battery Anode Grown by Atomic Layer Deposition. *RSC Adv.* 2016, *6*, 38024–38032.

286. Mahuli, N.; Saha, D.; Sarkar, S. K. Atomic Layer Deposition of P-Type Bi_2S_3. *J. Phys. Chem. C* 2017, *121*, 8136–8144.

287. Pickrahn, K. L.; Garg, A.; Bent, S. F. ALD of Ultrathin Ternary Oxide Electrocatalysts for Water Splitting. *ACS Catal.* 2015, *5*, 1609–1616.

288. Wang, H.; Wang, J. J.; Gordon, R.; Lehn, J. S. M.; Li, H.; Hong, D.; Shenai, D. V. Atomic Layer Deposition of Lanthanum-Based Ternary Oxides. *Electrochem. Solid State Lett.* 2009, *12*, G13–G15.

289. Heo, J.; Kim, S. B.; Gordon, R. G. Atomic Layer Deposited Zinc Tin Oxide Channel for Amorphous Oxide Thin Film Transistors. *Appl. Phys. Lett.* 2012, *101*, 113507–113512.

290. Wang, C.; Zhao, D.; Grice, C. R.; Liao, W.; Yu, Y.; Cimaroli, A.; Shrestha, N.; Roland, P. J.; Chen, J.; Yu, Z. et al. Low-Temperature Plasma-Enhanced Atomic Layer Deposition of Tin Oxide Electron Selective Layers for Highly Efficient Planar Perovskite Solar Cells. *J. Mater. Chem. A* 2016, *4*, 12080–12087.

291. Talkenberg, F.; Illhardt, S.; Radnoczi, G. Z.; Pecz, B.; Schmidl, G.; Schleusener, A.; Dikhanbayev, K.; Mussabek, G.; Gudovskikh, A.; Sivakov, V. Atomic Layer Deposition Precursor Step Repetition and Surface Plasma Pretreatment Influence on Semiconductor–Insulator–Semiconductor Heterojunction Solar Cell. *J. Vac. Sci. Technol. A* 2012, *30*, 021202–021211.

292. Sinha, S.; Nandib, D. K.; Kim, S. H.; Heo, J. Atomic-Layer-Deposited Buffer Layers for Thin Film Solar Cells Using Earth-Abundant Absorber Materials: A Review. *Sol. Energy Mater. Sol. Cells* 2018, *176*, 49–68.

293. Bugot, C.; Bouttemy, M.; Schneider, N.; Etcheberry, A.; Lincot, D.; Donsanti, F. New Insights on the Chemistry of Plasma-Enhanced Atomic Layer Deposition of Indium Oxysulfide Thin Films and Their Use as Buffer Layers in $Cu(In,Ga)Se_2$ Thin Film Solar Cell. *J. Vac. Sci. Technol. A* 2018, *36*, 061510–061520.

294. Binetti, S.; Garattini, P.; Mereu, R.; Le Donne, A.; Marchionna, S.; Gasparotto, A.; Meschia, M.; Pinus, I.; Acciarri, M. Fabricating $Cu(In,Ga)Se_2$ Solar Cells on Flexible Substrates by a New Roll-to-Roll Deposition System Suitable for Industrial Applications. *Semicond. Sci. Technol.* 2015, *30*, 105006–105014.

295. Klinkert, T. Comprehension and Optimisation of the Co-Evaporation Deposition of $Cu(In,Ga)Se_2$ Absorber Layers for Very High Efficiency Thin Film Solar Cells. 2015. https://tel.archives-ouvertes.fr/tel-01130052.

296. Chung, W. C.; Chang, M. B. Review of Catalysis and Plasma Performance on Dry Reforming of CH_4 and Possible Synergistic Effects. *Renew. Sustain. Energy Rev.* 2016, *62*, 13–31.

297. Pérez-Fortes, M.; Bocin-Dumitriu, A.; Tzimas, E. CO_2 Utilization Pathways: Techno-Economic Assessment and Market Opportunities. *Energy Procedia* 2014, *63*, 7968–7975.

298. Taherimeh, M.; Pescarmona, P. P. Green Polycarbonates Prepared by the Copolymerization of CO_2 with Epoxides. *J. Appl. Polym. Sci.* 2014, *131*, 41141–41158.

299. Affan, F. B. Direct Reaction of Carbon Dioxide to Polycarbonate. *J. Ecosys. Ecograph.* 2016, *6* (2), 1–11.

300. Federsel, C.; Jackstell, R.; Beller, M. State-of-the-Art Catalysts for Hydrogenation of Carbondioxide. *Angew. Chem. Int.* 2010, *49*, 6254–6261.

301. Huang, S. Y.; Liu, S. G.; Li, J. P.; Zhao, N.; Wei, W.; Sun, Y. H. Synthesis of Cyclic Carbonate from Carbon Dioxide and Diols over Metal Acetates. *J. Fuel Chem. Technol.* 2007, *35*, 701–706.

302. Büttner, H.; Longwitz, L.; Steinbauer, J.; Wulf, C.; Werner, T. Recent Developments in the Synthesis of Cyclic Carbonates from Epoxides and CO_2. *Top. Curr. Chem.* 2017, *375*, 50–56.

303. Zhao, X.; Sun, N.; Wang, S.; Li, F.; Wang, Y. Synthesis of Propylene Carbonate from Carbon Dioxide and 1,2-Propylene Glycol over Zinc Acetate Catalyst. *Ind. Eng. Chem. Res.* 2008, *47* (5), 1365–1369.

304. Li, H.; Niu, Y. High Yield Synthesis of Biodegradable Poly(Propylene Carbonate) from Carbon Dioxide and Propylene Oxide. *Polym. Adv. Technol.* 2016, *27* (9), 1191–1194.

305. Pérez-Fortes, M.; Schöneberger, J. C.; Boulamanti, A.; Tzimas, E. Methanol Synthesis Using Captured CO_2 as Raw Material: Techno-Economic and Environmental Assessment. *Appl. Energy* 2016, *161* (1), 718–732.

306. Kiss, A. A.; Pragt, J. J.; Vos, H. J.; Bargeman, V. G.; de Groot, T. Novel Efficient Process for Methanol Synthesis by CO_2 Hydrogenation. *Chem. Eng. J.* 2016, *284*, 260–269.

307. Chawl, S. K.; George, M.; PateL, F.; Patel, S. Production of Synthesis Gas by Carbon Dioxide Reforming of Methane over Nickel Based and Perovskite Catalysts. *Procedia Eng.* 2013, *51*, 461–466.

308. Sheng, W.; Kattel, S.; Yao, S.; Yan, B.; Liang, Z.; Hawxhurst, C. J.; Wuc, Q.; Chen, J. G. Electrochemical Reduction of CO_2 to Synthesis Gas with Controlled CO/H_2 Ratios. *Energy Environ. Sci.* 2017, *10*, 1180–1185.

309. Ross, M. B.; Dinh, C. T.; Li, Y.; Kim, D.; De Luna, P.; Sargent, E. H.; Yang, P. Tunable Cu Enrichment Enables Designer Syngas Electrosynthesis from CO_2. *J. Am. Chem. Soc.* 2017, *139* (7), 9359–9363.

310. Srinivas Ranjan, S.; Malik Sanjay, K.; Mahajani, M. Fischer-Tropsch Synthesis Using Bio-Syngas and CO_2. *Energy Sustain. Dev.* 2007, *11* (4), 66–71.

311. Yao, Y.; Liu, X.; Hildebrandt, D.; Glasser, D. Fischer–Tropsch Synthesis Using $H_2/CO/CO_2$ Syngas Mixtures: A Comparison of Paraffin to Olefin Ratios for Iron and Cobalt Based Catalysts. *Appl. Catal. A* 2012, *433–434*, 58–68.

312. Lillebø, A. L.; Holmen, A.; Enger, B. C.; Blekkan, E. A. *Fischer–Tropsch Conversion of Biomass-Derived Synthetic Gas to Liquid Fuels in Advances* in *Bioenergy: The Sustainability Challenge*; Lund, P., Byrne, J. A., Berndes, G., Vasalos, I.V. Eds.; Wiley, Chichester, UK, 2016.

313. Kim, H. H.; Teramoto, Y.; Ogata, A.; Takagi, H.; Nanba, T. Plasma Catalysis for Environmental Treatment and Energy Applications. *Plasma Chem. Plasma Process.* 2016, *36* (1), 45–72.

314. Somorjai, G. A. The Experimental Evidence of the Role of Surface Restructuring during Catalytic Reactions. *Catal. Lett.* 1992, *12* (1–3), 17–34.

315. Byun, Y.; Cho, M.; Hwang, S.; Chung, J. *Gasification for Practical Applications, Thermal Plasma Gasification of Municipal Solid Waste (MSW)*; IntechOpen, 2012; Vol. 1, Online version, DOI: 10.5772/48537.

316. Starikovskiy, A. Physics and Chemistry of Plasma-Assisted Combustion. *Philos. Trans. R. Soc. A* 2018, *373*, 20150074–20150082.

317. Ju, Y.; Sun, W. Plasma Assisted Combustion: Dynamics and Chemistry. *Prog. Energy Combust. Sci.* 2015, *48*, 21–83.

318. Starikovskiy, A.; Aleksandrov, N. Plasma-Assisted Ignition and Combustion. *Prog. Energy Combust. Sci.* 2013, *39* (1), 61–110.

319. Changming, D.; Chao, S.; Gong, X.; Ting, W.; Xiange, W. Plasma Methods for Metals Recovery from Metal–Containing Waste. *Waste Manage.* 2018, *77*, 373–387.

320. Song, H.; Hu, F.; Peng, Y.; Li, K.; Bai, S.; Li, J. Non-Thermal Plasma Catalysis for Chlorobenzene Removal over CoMn/TiO$_2$ and CeMn/TiO$_2$: Synergistic Effect of Chemical Catalysis and Dielectric Constant. *Chem. Eng. J.* 2018, *347*, 447–454.

321. Aymen, A.; Guiza, M.; Bouzaza, A.; Aboussaoud, W.; Soutrel, I.; Ouederni, A.; Wolbert, D.; Rtimi, S. Abatement of Ammonia and Butyraldehyde under Non-Thermal Plasma and Photocatalysis: Oxidation Processes for the Removal of Mixture Pollutants at Pilot Scale. *Chem. Eng. J.* 2018, *344*, 165–172.

322. Punčochářa, M.; Rujb, B.; Chatterj, P. K. Development of Process for Disposal of Plastic Waste Using Plasma Pyrolysis Technology and Option for Energy Recovery. *Procedia Eng.* 2012, *42*, 420–430.

323. Tang, L.; Huang, H.; Hao, H.; Zhao, K. Development of Plasma Pyrolysis/Gasification Systems for Energy Efficient and Environmentally Sound Waste Disposal. *J. Electrost.* 2013, *71* (5), 839–847.

324. Jin, L.; Li, Y.; Feng, Y.; Hua, H.; Zhu, A. Integrated Process of Coal Pyrolysis with CO_2 Reforming of Methane by Spark Discharge Plasma. *J. Anal. Appl. Pyrolysis* 2017, *126*, 194–200.

325. Sun, D. L.; Wang, F.; Hong, R. Y.; Xie, C. R. Preparation of Carbon Black via Arc Discharge Plasma Enhanced by Thermal Pyrolysis. *Diam. Relat. Mater.* 2016, *61*, 21–31.

326. Ma, J.; Su, B.; Wen, G.; Yang, Q.; Ren, Q.; Yang, Y.; Xing, H. Pyrolysis of Pulverized Coal to Acetylene in Magnetically Rotating Hydrogen Plasma Reactor. *Fuel Proc. Technol.* 2017, *167*, 721–729.

327. Lian, H. Y.; Li, X. S.; Liu, J. L.; Zhu, X.; Zhu, A. M. Oxidative Pyrolysis Reforming of Methanol in Warm Plasma for an On-Board Hydrogen Production. *Int. J. Hydrog. Energy* 2017, *42* (19), 13617–13624.

328. Perna, A.; Minutillo, M.; Lubrano Lavadera, A.; Jannelli, E. Combining Plasma Gasification and Solid Oxide Cell Technologies in Advanced Power Plants for Waste to Energy and Electric Energy Storage Applications. *Waste Manage.* 2018, *73*, 424–438.

329. Schultz P.G. Commentary on Combinatorial Chemistry. *Appl. Catal. A* 2003, *254* (1), 3–4.

330. Potyrailo, R.; Rajan, K.; Stoewe, K.; Takeuchi, I.; Chisholm, B.; Lam, H. Combinatorial and High-Throughput Screening of Materials Libraries: Review of State of the Art. *ACS Comb. Sci.* 2011, *13* (6), 579–633. https://doi.org/10.1021/co200007w.

331. Takeuchi, I.; Lauterbach, J.; Fasolka, M. J. Combinatorial Materials Synthesis. *Mater. Today* 2005, *8* (10), 18–26.

332. McFarland, E. W.; Weinberg, W. H. Combinatorial Approaches to Materials Discovery. *Trends Biotechnol.* 1999, *17* (3), 107–115. https://doi.org/10.1016/S0167-7799(98)01275-X.

333. Liu, Y.; Padmanabhan, J.; Cheung, B.; Liu, J.; Chen, Z.; Scanley, B. E.; Wesolowski, D.; Pressley, M.; Broadbridge, C. C.; Altman, S. et al. Combinatorial Development of Antibacterial Zr-Cu-Al-Ag Thin Film Metallic Glasses. *Sci. Rep.* 2016, *6* (1). https://doi.org/10.1038/srep26950.

334. Koinuma, H.; Takeuchi, I. Combinatorial Solid-State Chemistry of Inorganic Materials. *Nat. Mater.* 2004, *3* (7), 429–438. https://doi.org/10.1038/nmat1157.

335. Jandeleit, B.; Schaefer, D. J.; Powers, T. S.; Turner, H. W.; Weinberg, W. H. Combinatorial Materials Science and Catalysis. *Angew. Chem. Int. Ed.* 1999, 38 (17), 2494–2532.

336. Gebhard, M.; Mitschker, F.; Hoppe, C.; Aghaee, M.; Rogalla, D.; Creatore, M.; Grundmeier, G.; Awakowicz, P.; Devi, A. A Combinatorial Approach to Enhance Barrier Properties of Thin Films on Polymers: Seeding and Capping of PECVD Thin Films by PEALD. *Plasma Process. Polym.* 2018, *15* (5), 1700209. https://doi.org/10.1002/ppap.201700209.

337. Ding, S.; Liu, Y.; Li, Y.; Liu, Z.; Sohn, S.; Walker, F. J.; Schroers, J. Combinatorial Development of Bulk Metallic Glasses. *Nat. Mater.* 2014, *13* (5), 494–500. https://doi.org/10.1038/nmat3939.

338. Carson Meredith, J.; Karim, A.; Amis, E. J. Combinatorial Methods for Investigations in Polymer Materials Science. *MRS Bull.* 2002, *27* (04), 330–335. https://doi.org/10.1557/mrs2002.101.

339. Xiang, X. D.; Takeuchi, I. *Combinatorial Materials Synthesis*; Marcel Dekker Inc., New York, USA, 2003.

340. Potyrailo, R. A.; Maier, W. F. *Combinatorial and High-Throughput Discovery and Optimization of Catalysts and Materials*; CRC Press, Boca Raton, FL, 2006.

341. Malhotra, R. *Combinatorial Approaches to Materials Development*; American Chemical Society, Washington, DC, 2002; Vol. 814.

342. Hanak, J. J. The "Multiple-Sample Concept" in Materials Research: Synthesis, Compositional Analysis and Testing of Entire Multicomponent Systems. *J. Mater. Sci.* 1970, 5 (11), 964–971.

343. Xiang, X.-D.; Sun, X.; Briceno, G.; Lou, Y.; Wang, K.-A.; Chang, H.; Wallace-Freedman, W. G.; Chen, S.-W.; Schultz, P. G. A Combinatorial Approach to Materials Discovery. *Science* 1995, *268* (5218), 1738–1740. https://doi.org/10.1126/science.268.5218.1738.

344. Potyrailo, R. A.; Amis, E. J. *High Throughput Analysis: A Tool for Combinatorial Materials Science*; Kluwer Academic/Plenum, New York, NY, 2003.

345. Potyrailo, R. A.; Takeuchi, I. Combinatorial and High-Throughput Materials Research. *Meas. Sci. Technol.* 2005, *16* (1), 1–4.

346. Kennedy, K.; Stefansky, T.; Davy, G.; Zackay, V. F.; Parker, E. R. Rapid Method for Determining Ternary-Alloy Phase Diagrams. *J. Appl. Phys.* 1965, *36*, 3808–3810.

347. Hasegawa, K.; Ahmet, P.; Okazaki, N.; Hasegawa, T.; Fujimoto, K.; Watanabe, M.; Chikyow, T.; Koinuma, H. Amorphous Stability of HfO_2 Based Ternary and Binary Composition Spread Oxide Films as Alternative Gate Dielectrics. *Appl. Surf. Sci.* 2004, *223* (1–3), 229–232.

348. Chikyow, T.; Ohmori, K.; Nagata, T.; Umezawa, N.; Haemori, M.; Yoshitake, M.; Hasegawa, T.; Koinuma, H.; Yamada, K. Landscape of Combinatorial Materials Exploration and High Throughput Characterizations for the Post-CMOS Devices. *IEEE* 2008, 66–67.

349. Chikyow, T.; Hasegawa, K.; Tamori, T.; Ohmori, K.; Umezawa, K.; Nakajima, K.; Yamada, K.; Koinuma, H. Combinatorial Materials Exploration and Composition Tuning for the Future Gate Stack Structure. *Proc. 8th Int. Conf. Solid-State Integr. Circuit Technol.* 2006, 6.

350. Klamo, J. L.; Schenck, P. K.; Burke, P. G.; Chang, K.-S.; Green, M. L. Manipulation of the Crystallinity Boundary of Pulsed Laser Deposited High-k HfO_2–TiO_2–Y_2O_3 Combinatorial Thin Films. *J. Appl. Phys.* 2010, *107* (5), 054101. https://doi.org/10.1063/1.3294607.

351. Hubbard, K. J.; Schlom, D. G. Thermodynamic Stability of Binary Oxides in Contact with Silicon. *J. Mater. Res.* 1996, *11* (11), 2757–2776.

352. Ritala, M.; Leskela, M. Deposition and Processing of Thin Films. In *Handbook of Thin Film Materials*; Academic Press, San Diego, 2002; Vol. 1.

353. Ohmori, K.; Chikyow, T.; Hosoi, T.; Watanabe, H.; Nakajima, K.; Adachi, T.; Ishikawa, A.; Sugita, Y.; Nara, Y.; Ohji, Y. et al. Wide Controllability of Flatband Voltage by Tuning Crystalline Microstructures in Metal Gate Electrodes. *Proc. Int. Electron Devices Meet. IEDM 2007* 2007, 345–348.

354. Chang, K. S.; Green, M. L.; Suehle, J.; Vogel, E. M.; Xiong, H.; Hattrick-Simpers, J.; Takeuchi, I.; Famodu, O.; Ohmori, K.; Ahmet, P. et al. Combinatorial Study of Ni-Ti–Pt Ternary Metal Gate Electrodes on HfO_2 for the Advanced Gate Stack. *Appl. Phys. Lett.* 2006, *89* (14), 142108–142115.

355. Ohmori, K.; Ahmet, P.; Shiraishi, K.; Yamabe, K.; Watanabe, H.; Akasaka, Y.; Umezawa, N.; Nakajima, K.; Yoshitake, M.; Nakayama, T. et al. Controllability of Flatband Voltage in High-k Gate Stack Structures – Remarkable Advantages of La_2O_3 over HfO_2. *Proc. 8th Int. Conf. Solid-State Integr. Circuit Technol.* 2006, 376–379.

356. Huang, C.; Wang, M.; Deng, Z.; Cao, Y.; Liu, Q.; Huang, Z.; Liu, Y.; Guo, W.; Huang, Q. Low Content Indium-Doped Zinc Oxide Films with Tunable Work Function Fabricated through Magnetron Sputtering. *Semicond. Sci. Technol.* 2010, *25* (4), 045008.

357. Bai, W. P.; Bae, S. H.; Wen, H. C.; Mathew, S.; Bera, L. K.; Balasubramanian, N.; Yamada, N.; Li, M. F.; Kwong, D. L. Three-Layer Laminated Metal Gate Electrodes with Tunable Work Functions for CMOS Applications. *IEEE Electron Device Lett.* 2005, *26* (4), 231–233.

358. Choi, K.; Alshareef, H. N.; Wen, H. C.; Harris, H.; Luan, H.; Senzaki, Y.; Lysaght, P.; Majhi, P.; Lee, B. H. Effective Work Function Modification of Atomic-Layer-Deposited-TaN Film by Capping Layer. *Appl. Phys. Lett.* 2006, *89* (3), 032113.

359. Fukumura, T.; Ohtani, M.; Kawasaki, M.; Okimoto, Y.; Kageyama, T.; Koida, T.; Hasegawa, T.; Tokura, T.; Koinuma, H. Rapid Construction of a Phase Diagram of Doped Mott Insulators with a Composition-Spread Approach. *Appl. Phys. Lett.* 2000, *77* (21), 3426–3428.

360. Yoo, Y. K.; Duewer, F.; Fukumura, T.; Yang, H. T.; Yi, D.; Liu, S.; Chang, H. S.; Hasegawa, T.; Kawasaki, M.; Koinuma, H. et al. Strong Correlation between High-Temperature Electronic and Low-Temperature Magnetic Ordering in $La_{1-x}Ca_xMnO_3$ Continuous Phase Diagram. *Phys. Rev.* 2001, *B63* (22), 224421.

361. Turchinskaya, M. J.; Bendersky, L. A.; Shapiro, A. J.; Chang, K. S.; Takeuchi, I.; Roytburd, A. L. Rapid Constructing Magnetic Phase Diagrams by Magneto-Optical Imaging of Composition Spread Films. *J. Mater. Res.* 2004, *19* (9), 2546–2548.

362. Iwasaki, Y.; Fukumura, T.; Kimura, H.; Ohkubo, A.; Hasegawa, T.; Hirose, Y.; Makino, T.; Ueno, K.; Kawasaki, M. High-Throughput Screening of Ultraviolet–Visible Magnetooptical Properties of Spinel Ferrite $(Zn,Co)Fe_2O_4$ Solid Solution Epitaxial Film by a Composition-Spread Approach. *Appl. Phys. Express* 2010, *3* (10), 103001.

363. Hanak, J. J.; Gittleman, J. I. Iron-Nickel-Silica Ferromagnetic Cermets. *Proc. AIP* 1972, *10*, 961–965.

364. Yoo, Y. K.; Xue, Q. Z.; Chu, Y. S.; Xu, S. F.; Hangen, U.; Lee, H. C.; Stein, H. C.; Xiang, X. D. Identification of Amorphous Phases in the Fe–Ni–Co Ternary Alloy System Using Continuous Phase Diagram Material Chips. *Intermetallics* 2006, *14* (3), 241–247.

365. Muduli, P. K.; Rice, W. C.; He, L.; Collins, B. A.; Chu, Y. S.; Tsui, F. Study of Magnetic Anisotropy and Magnetization Reversal Using the Quadratic Magnetooptical Effect in Epitaxial $Co_xMn_yGe_z(111)$ Films. *J. Phys. Condens. Matter.* 2009, *21* (29), 296005.

366. Degroot, R. A.; Mueller, F. M.; Vanengen, P. G.; Buschow, K. H. J. New Class of Materials: Half-Metallic Ferromagnets. *Phys. Rev. Lett.* 1983, *50* (25), 2024–2027.

367. Takeuchi, I.; Famodu, O. O.; Read, J. C.; Aronova, M. A.; Chang, K.-S.; Craciunescu, C.; Lofland, S. E.; Wuttig, M.; Wellstood, F. C.; Knauss, L. et al. Identification of Novel Compositions of Ferromagnetic Shape-Memory Alloys Using Composition Spreads. *Nat. Mater.* 2003, *2* (3), 180–184. https://doi.org/10.1038/nmat829.

368. Hanak, J. J. *Combinatorial Materials Synthesis*; Xiang, X.-D., Takeuchi, I. Eds.; Dekker, New York, 2003.

369. Jin, K.; Suchoski, R.; Fackler, S.; Zhang, Y.; Pan, X.; Greene, R. L.; Takeuchi, I. Combinatorial Search of Superconductivity in Fe-B Composition Spreads. *APL Mater.* 2013, *1* (4), 042101. https://doi.org/10.1063/1.4822435.

370. Clayhold, J. A.; Kerns, B. M.; Schroer, M. D.; Rench, D. W.; Logvenov, G.; Bollinger, A. T.; Bozovic, I. Thermal Properties of Solids at Room and Cryogenic Temperatures. *Rev. Sci. Instrum.* 2008, *79* (3), 093906.

371. Wang, J.; Yoo, Y.; Takeuchi, I.; Sun, X.; Chang, H.; Xiang, X.; Schultz, P. G. Identification of a Blue Photoluminescent Composite Material from a Combinatorial Library. *Science* 1998, *279* (5357), 1712–1714.

372. Kanai, Y. Electrical Properties of Indium- Tin-Oxide Single Crystals. *Jpn. J. Appl. Phys.* 1984, *23*, L12.

373. Taylor, M. P.; Readey, D. W.; Teplin, C. W.; van Hest, M.; Alleman, J. L.; Dabney, M. S.; Gedvilas, L. M.; Keyes, B. M.; To, B.; Parilla, P. A. et al. Combinatorial Growth and Analysis of the Transparent Conducting Oxide ZnO/In (IZO). *Macromol. Rapid Commun.* 2004, *25* (1), 344–347.

374. van Hest, M.; Dabney, M. S.; Perkins, J. D.; Ginley, D. S.; Taylor, M. P. Titanium-Doped Indium Oxide: A High-Mobility Transparent Conductor. *Appl. Phys. Lett.* 2005, *87* (3), 032111.

375. Taylor, M. P.; Readey, D. W.; Teplin, C. W.; van Hest, M.; Alleman, J. L.; Dabney, M. S.; Gedvilas, L. M.; Keyes, B. M.; To, B.; Perkins, J. D. et al. The Electrical, Optical and Structural Properties of $In_xZn_{1-x}O_y$ ($0 \leq x \leq 1$) Thin Films by Combinatorial Techniques. *Meas. Sci. Technol.* 2005, *16* (1), 90–94.

376. Perkins, J. D.; Taylor, M. P.; van Hest, M.; Teplin, C. W.; Alleman, J. L.; Dabney, M. S.; Gedvilas, L. M.; Keyes, B. M.; To, B.; Readey, D. W. et al. Combinatorial Optimization of Transparent Conducting Oxides (TCOs) for PV. *Proc. Conf. Rec. Thirty-First IEEE Photovolt. Spec. Conf.* 2005, 145–147.

377. Koida, T.; Kondo, M. Comparative Studies of Transparent Conductive Ti-, Zr-, and Sn-Doped In_2O_3 Using a Combinatorial Approach. *J. Appl. Phys.* 2007, *101* (6), 063713.

378. Sheel, D. W.; Yates, H. M.; Evans, R.; Dagkaldiran, U.; Gordijn, A.; Finger, Z.; Remes, Z.; Vanecek, M. Atmospheric Pressure Chemical Vapour Deposition of F Doped SnO_2 for Optimum Performance Solar Cells. *Thin Solid Films* 2009, *517* (10), 3061–3065.

379. Bhachu, D. S.; Waugh, M. R.; Zeissler, K.; Branford, W. R.; Parkin, I. P. Textured Fluorine-Doped Tin Dioxide Films Formed by Chemical Vapour Deposition. *Chem. Eur. J.* 2011, *17* (41), 11613–11621. https://doi.org/10.1002/chem.201100399.

380. Jung, K.; Choi, W.-K.; Yoon, S.-J.; Kim, H. J.; Choi, J.-W. Electrical and Optical Properties of Ga Doped Zinc Oxide Thin Films Deposited at Room Temperature by Continuous Composition Spread. *Appl. Surf. Sci.* 2010, *256* (21), 6219–6223. https://doi.org/10.1016/j.apsusc.2010.03.144.

381. Gorrie, C. W.; Reese, M.; Perkins, J. D.; van Hest, M.; Alleman, J. L.; Dabney, M. S.; To, B.; Ginley, D. S.; Berry, J. J. Proceedings of the 33rd IEEE Photovoltaic Specialists Conference (IEEE, New York, 2008), Vol. 1–4. 2008, *1–4*, 635–637.

382. Wang, Q.; Perkins, J. D.; Branz, H. M.; Alleman, J. L.; Duncan, C.; Ginley, D. S. Combinatorial Synthesis of Solid State Electronic Materials for Renewable Energy Applications. *Appl. Surf. Sci.* 2002, *189* (3–4), 271–276.

383. Zakutayev, A.; Perkins, J. D.; Parilla, P. A.; Widjonarko, N. E.; Sigdel, A. K.; Berry, J. J.; Ginley, D. S. Zn–Ni–Co–O Wide-Band-Gap p-Type Conductive Oxides with High Work Functions. *MRS Commun.* 2011, *1* (1), 23–26.

384. Belosludov, R.; Cheettu, A. S. S.; Inaba, Y.; Oumi, Y.; Takami, S.; Kubo, M.; Miyamoto, A. et al. Combinatorial Computational Chemistry Approach to the Design of Metal Oxide Electronics Materials. *Proc. SPIE- Int. Soc. Opt. Eng.* 2000, 3941, 2–10.

385. Takada, K.; Fujimoto, K.; Sasaki, T.; Watanabe, M. Combinatorial Electrode Array for High-Throughput Evaluation of Combinatorial Library for Electrode Materials. *Appl. Surf. Sci.* 2004, *223* (1–3), 210–213.

386. Ceder, G. Opportunities and Challenges for First-Principles Materials Design and Applications to Li Battery Materials. *MRS Bull.* 2010, *35* (9), 639–701.

387. Fujimoto, K.; Onoda, K.; Ito, S. Exploration of Layered-Type Pseudo Four-Component Li–Ni–Co–Ti Oxides. *Appl. Surf. Sci.* 2007, *254* (3), 704–708.

388. Fujimoto, K.; Takada, K.; Sasaki, T.; Watanabe, M. Combinatorial Approach for Powder Preparation of Pseudo-Ternary System $LiO_{0.5}-X-TiO_2$ (X: $FeO_{1.5}$, $CrO_{1.5}$ and NiO). *Appl. Surf. Sci.* 2004, *233* (1–3), 49–53.

389. Roberts, M.; Owen, J. High-Throughput Method to Study the Effect of Precursors and Temperature, Applied to the Synthesis of $LiNi_{1/3}Co_{1/3}Mn_{1/3}O_2$ for Lithium Batteries. *ACS Comb. Sci.* 2011, *16* (2), 126–134.

390. Whitacre, J. F.; West, W. C.; Ratnakumar, B. V. A Combinatorial Study of $Li_y Mn_x Ni_{2-x}O_4$ Cathode Materials Using Microfabricated Solid-State Electrochemical Cells. *J. Electrochem. Soc.* 2003, *150* (12), A1676–A1683.

391. Todd, A. D. W.; Mar, R. E.; Dahn, J. R. Combinatorial Study of Tin-Transition Metal Alloys as Negative Electrodes for Lithium-Ion Batteries. *J. Electrochem. Soc.* 2006, *153* (10), A1998–A2005.

392. Todd, A. D. W.; Mar, R. E.; Dahn, J. R. Tin–Transition Metal–Carbon Systems for Lithium-Ion Battery Negative Electrodes. *J. Electrochem. Soc.* 2007, *154* (6), A597–A604.

393. Sun, Z. B.; Wang, X. D.; Li, X. P.; Zhao, M. S.; Li, Y.; Zhu, Y. M.; Song, X. P. Electrochemical Properties of Melt-Spun Al–Si–Mn Alloy Anodes for Lithium-Ion Batteries. *J. Power Sources* 2008, *182* (1), 353–358.

394. Fleischauer, M. D.; Dahn, J. R. Combinatorial Investigations of the Si-Al-Mn System for Li-Ion Battery Applications. *J. Electrochem. Soc. 151* (8), A1216–A1221.

395. Minami, H.; Itaka, K.; Kawaji, H.; Wang, Q. J.; Koinuma, H.; Lippmaa, M. Rapid Synthesis and Characterization of (Ca1–xBax)3Co4O9 Thin Films Using Combinatorial Methods. *Appl. Surf. Sci.* 2002, *197*, 442–447.

396. Otani, M.; Lowhorn, N. D.; Schenck, P. K.; Wong-Ng, W.; Green, M. L.; Itaka, K.; Koinuma, H. A High-Throughput Thermoelectric Power-Factor Screening Tool for Rapid Construction of Thermoelectric Property Diagrams. *Appl. Phys. Lett.* 2007, *91*, 132102.

397. Otani, M.; Thomas, E. L.; Wong-Ng, W.; Schenck, P. K.; Chang, K. S.; Lowhorn, N. D.; Green, M. L.; Ohguchi, H. A High-Throughput Screening System for Thermoelectric Material Exploration Based on a Combinatorial Film Approach. *Jpn. J. Appl. Phys.* 2009, *48* (5), 05EB02.

398. Watanabe, M.; Kita, T.; Fukumura, T.; Ohtomo, A.; Ueno, K.; Kawasaki, M. High-Throughput Screening for Combinatorial Thin-Film Library of Thermoelectric Materials. *J. Comb. Chem.* 2008, *10* (2), 175–178.

399. Yamamoto, A.; Obara, H.; Ueno, K. Optimization of Thermoelectric Properties of Ni-Cu Based Alloy Through Combinatorial Approach. In *Thermoelectric Power Generation*; Ed. Hogan, T. P., Yang, J., Funahashi, R., Tritt, T. M., Cambridge University Press, New York, USA, 2008.

400. Kojima, Y.; Miyaoka, H.; Ichikawa, T. *Chap.* 5 in *Batteries, Hydrogen Storage and Fuel Cells*; Suib, S. Ed.; Elsevier, 2013.

401. Sakintuna, B.; Darkrim, F. L.; Hirscher, M. Metal Hydride Materials for Solid Hydrogen Storage. A Review. *Int. J. Hydrog. Energy* 2007, *32*, 1121–1140.

402. Zaluski, L.; Zaluska, A.; Ström-Olsen, J. O. Nanocrystalline Metal Hydrides. *J. Alloy. Compd.* 1997, *253–254*, 70–79.

403. Zaluski, L.; Zaluska, A.; Ström-Olsen, J. O. Hydrogen Absorption in Nanocrystalline Mg2Ni Formed by Mechanical Alloying. *J. Alloy. Compd.* 1995, *217*, 245–249.

404. Holtz, R. L.; Imam, M. A. Hydrogen Storage Characteristics of Ballmilled Magnesium–Nickel and Magnesium–Iron Alloys. *J. Mater. Sci.* 1999, *34*, 2655–2663.

405. Iwakura, C.; Nohara, S.; Zhang, S. G.; Inoue, H. Hydriding and Dehydriding Characteristics of an Amorphous Mg_2Ni–Ni Composite. *J. Alloy. Compd.* 1999, *285*, 246–249.

406. Meisner, G. P.; Scullin, M. L.; Balogh, M. P.; Pinkerton, F. E.; Meyer, M. S. Hydrogen Release from Mixtures of Lithium Borohydride and Lithium Amide: A Phase Diagram Study. *J. Phys. Chem. B* 2006, *110* (9), 4186–4192.

407. Lu, J.; Fang, Z. Z.; Choi, Y. J.; Sohn, H. Y. Potential of Binary Lithium Magnesium Nitride for Hydrogen Storage Applications. *J. Phys. Chem. C* 2007, *111* (32), 12129–12134.

408. Hattrick-Simpers, R.; Maslar, J. E.; Niemann, M. U.; Chiu, C.; Srinivasan, S. S.; Stefanakos, E. K.; Bendersky, L. A. Raman Spectroscopic Observation of Dehydrogenation in Ball-Milled $LiNH_2$–$LiBH_4$–MgH_2 Nanoparticles. *Int. J. Hydrog. Energy* 2010, *35* (12), 6323–6331.

409. Yang, J.; Sudik, A.; Siegel, D. J.; Halliday, D.; Drews, A.; Carter, R. O.; Wolverton, C.; Lewis, G. J.; Sachtler, J. W. A.; Low, J. J. et al. Improved Dehydrogenation Cycle Performance of the $1.1MgH_2$–$2LiNH_2$–$0.1LiBH_4$ System by Addition of $LaNi_{4.5}Mn_{0.5}$ Alloy. *Angew. Chem. Int. Ed.* 2008, *47* (5), 882–887.

410. Oguchi, H.; Heilweil, E. J.; Josell, D.; Bendersky, L. A. Infrared Emission Imaging as a Tool for Characterization of Hydrogen Storage Materials. *J. Alloy. Compd.* 2009, *477* (1–2), 8–15.

411. Olk, C. H.; Tibbetts, G. G.; Simon, D.; Moleski, J. J. Combinatorial Preparation and Infrared Screening of Hydrogen Sorbing Metal Alloys. *J. Appl. Phys.* 2003, *94* (1), 720–725.

412. Garcia, G.; Domenech-Ferrer, R.; Pi, F.; Santiso, J.; Rodriguez-Viejo, J. Combinatorial Synthesis and Hydrogenation of Mg/Al Libraries Prepared by Electron Beam Physical Vapor Deposition. *J. Comb. Chem.* 2007, *9* (2), 230–236.

413. Gremaud, R.; Broedersz, C. P.; Borsa, D. M.; Borgschulte, A.; Mauron, P.; Schreuders, H.; Rector, J. H.; Dam, B.; Griessen, R. Hydrogenography: An Optical Combinatorial Method To Find New Light-Weight Hydrogen-Storage Materials. *Adv. Mater.* 2007, *19* (19), 2813–2817.

414. Gremaud, R.; Broedersz, C. P.; Borgschulte, A.; van Setten, M. J.; Schreuders, H.; Slaman, M.; Dama, B.; Griessen, R. Hydrogenography of MgyNi1–yHx Gradient Thin Films: Interplay between the Thermodynamics and Kinetics of Hydrogenation. *Acta Mater.* 2010, *58* (2), 658–668.

415. Borgschulte, A.; Chacon, C.; van Mechelen, J. L. M.; Schreuders, H.; Zuttel, A.; Hjorvarsson, B.; Dam, B.; Griessen, R. Structural and Optical Properties of MgxAl1-XHy Gradient Thin Films: A Combinatorial Approach. *Appl. Phys. A* 2006, *84* (1–2), 77–85.

416. Gremaud, R.; Borgschulte, A.; Lohstroh, W.; Schreuders, H.; Zuttel, A.; Dam, B.; Griessen, R. Ti-Catalyzed Mg(AlH$_4$)$_2$: A Reversible Hydrogen Storage Material. *J. Alloy. Compd.* 2005, *404*, 775–778.

417. Broedersz, C. P.; Gremaud, R.; Dam, B.; Griessen, R.; Lovvik, O. M. Highly Destabilized Mg-Ti-Ni-H System Investigated by Density Functional Theory and Hydrogenography. *Phys. Rev. B* 2008, *77* (2), 024204–024214.

418. Ludwig, A.; Cao, J.; Savan, A.; Ehmann, M. High-Throughput Characterization of Hydrogen Storage Materials Using Thin Films on Micromachined Si Substrates. *J. Alloy. Compd.* 2007, *446*, 516–521.

419. Ludwig, A.; Cao, J.; Dam, B.; Gremaud, R. Opto-Mechanical Characterization of Hydrogen Storage Properties of Mg–Ni Thin Film Composition Spreads. *Appl. Surf. Sci.* 2007, *254* (3), 682–686.

420. Park, J. B.; Bronzino, J. D. *Biomaterials: Principles and Applications*; CRC Press, Boca Raton, FL, 2003.

421. Ratner, B. D.; Hoffman, A. S.; Schoen, F. J.; Lemons, J. E. *Biomaterials Science. An Introduction to Materials*; Elsevier, San Diego, CA, 2004.

422. Standard ISO 10993-1: Biological Evaluation of Medical Devices – Part 1: Evaluation and Testing within a Risk Management Process. International Organization for Standardization 2016.

423. Simon, C. G.; Yang, Y.; Thomas, V.; Dorsey, S. M.; Morgan, A. W. Cell Interactions with Biomaterials Gradients and Arrays. *Comb. Chem. High Throughput Screen.* 2009, *12* (6), 544–553.

424. Simon, C. G.; Lin-Gibson, S. Combinatorial and High-Throughput Screening of Biomaterials. *Adv. Mater.* 2011, *23* (3), 369–387. https://doi.org/10.1002/adma.201001763.

425. Liu, E.; Treiser, M. D.; Patel, H.; Sung, H. J.; Roskov, K. E.; Kohn, J.; Becker, M. L.; Moghe, P. V. High-Content Profiling of Cell Responsiveness to Graded Substrates Based on Combinatorially Variant Polymers. *Comb. Chem. High Throughput Screen.* 2009, *12*(7), 646–655.

426. Jäger, M.; Zilkens, K.; Krauspe, R. Significance of Nano- and Microtopography for Cell-Surface Interactions in Orthopaedic Implants. *J. Biomed. Biotechnol.* 2007, *8*, 69036–69055.

427. Horbett, T. A.; Brash, J. L. *Proteins at Interfaces: An Overview in Proteins at the Interfaces II: Fundamentals and Applications*; ACS Symposium; American Chemical Society, Washington, DC, 1995; Vol. 602.

428. Carré, A.; Lacarrière, V. How Substrate Properties Control Cell Adhesion. A Physical–Chemical Approach. *J. Adhes. Sci. Technol.* 2010, *24* (5), 815–830.

429. Simon, C. G.; Eidelmanb, N.; Kennedy, S. B.; Sehgal, A.; Khatri, C. A.; Washburn, N. R. Combinatorial Screening of Cell Proliferation on Poly(L-Lactic Acid)/Poly(D,L-Lactic Acid) Blends. *Biomaterials* 2005, *26*, 6906–6915.

430. Kennedy, S. B.; Washburn, N. R.; Simon, C. G.; Amis, E. J. Combinatorial Screen of the Effect of Surface Energy on Fibronectin-Mediated Osteoblast Adhesion, Spreading and Proliferation. *Biomaterials* 2006, *27* (20), 3817–3824.

431. Mei, Y.; Wu, T.; Xu, C.; Langenbach, K. J.; Elliott, J. T.; Vogt, B. D.; Beers, K. L.; Amis, E. J.; Washburn, N. R. Tuning Cell Adhesion on Gradient Poly(2-Hydroxyethyl Methacrylate)-Grafted Surfaces. *Langmuir* 2005, *21* (26), 12309–12314.

432. Usprech, J.; Romero, D. A.; Amon, C. H.; Simmons, C. A. Combinatorial Screening of 3D Biomaterial Properties That Promote Myofibrogenesis for Mesenchymal Stromal Cell-Based Heart Valve Tissue Engineering. *Acta Biomater.* 2017, *58*, 34–43.

433. Zhou, Q.; Ocampo, O. C.; Guimarães, C. F.; Kühn, P. T.; van Kooten, T. G.; van Rijn, P. Screening Platform for Cell Contact Guidance Based on Inorganic Biomaterial Micro/Nanotopographical Gradients. *ACS Appl. Mater. Interfaces* 2017, *9* (37), 31433–31445.

434. Floren, M.; Tan, W. Three-Dimensional, Soft Neotissue Arrays as High Throughput Platforms for the Interrogation of Engineered Tissue Environments. *Biomaterials* 2015, *59*, 39–52.

435. Dolatshahi-Pirouz, A.; Nikkhah, M.; Gaharwar, A. K.; Hashmi, B.; Guermani, E.; Aliabadi, H.; Camci-Unal, G.; Ferrante, T.; Foss, M.; Ingber, D. E. et al. A Combinatorial Cell-Laden Gel Microarray for Inducing Osteogenic Differentiation of Human Mesenchymal Stem Cells. *Sci. Rep.* 2014, *4*, 3896–3905.

436. Mendes, L. F.; Tam, W. L.; Chai, Y. C.; Geris, L.; Luyten, F. P.; Roberts, S. J. Combinatorial Analysis of Growth Factors Reveals the Contribution of Bone Morphogenetic Proteins to Chondrogenic Differentiation of Human Periosteal Cells. *Tissue Eng. C* 2016, *22* (5), 473–486.

437. Wu, X.; Schultz, P. G. Synthesis at the Interface of Chemistry and Biology. *J. Am. Chem. Soc.* 2009, *131*, 12497–12515.

438. Dolle, R. E.; Le Bourdonnec, B.; Goodman, A. J.; Morales, G. A.; Thomas, C. J.; Zhang, W. Comprehensive Survey of Chemical Libraries for Drug Discovery and Chemical Biology. *J. Comb. Chem.* 2009, *11* (5), 739–790.

439. Meredith, J. C. Advances in Combinatorial and High-Throughput Screening of Biofunctional Polymers for Gene Delivery, Tissue Engineering and Anti-Fouling Coatings. *J. Mater. Chem.* 2009, *19* (1), 34–45.

440. Zelikin, A. N. Drug Releasing Polymer Thin Films: New Era of Surface-Mediated Drug Delivery. *ACS Nano* 2010, *4* (5), 2494–2509.

441. Andersen, A. H. F.; Riber, C. F.; Zuwala, K.; Tolstrup, M.; Dagnæs-Hansen, F.; Denton, P. W.; Zelikin, A. N. Long-Acting, Potent Delivery of Combination Antiretroviral Therapy. *ACS Macro Lett.*, 2018, *7* (5), 587–591.

Index

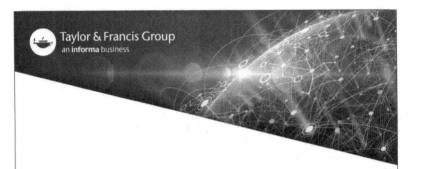